AF790725

Biosensors for Virus Detection

IOP Series in Sensors and Sensor Systems

The IOP Series in Sensors and Sensor Systems includes books on all aspects of the science and technology of sensors and sensor systems. Spanning fundamentals, fabrication, applications and processing, the series aims to provide a library for instrument and measurement scientists, engineers and technologists in universities and industry.

The series seeks (but is not restricted to) publications in the following topics:

- Advanced materials for sensing
- Biosensors
- Chemical sensors
- Industrial applications
- The Internet of Things
- Lab-on-a-chip
- Localization and object tracking
- Manufacturing and packaging
- Mechanisms, modelling and simulations
- MEMS/NEMS
- Micro- and nanosensors
- Non-destructive testing
- Optoelectronic and photonic sensors
- Optomechanical sensors
- Physical sensors
- Remote sensors
- Sensing for health, safety and security
- Sensing principles
- Sensing systems
- Sensor arrays
- Sensor devices
- Sensor networks
- Sensor technology and applications
- Signal processing and data analysis
- Smart sensors and monitoring
- Telemetry

Authors are encouraged to take advantage of electronic publication through the use of colour, animations, video and interactive elements to enhance the reader experience.

Biosensors for Virus Detection

Edited by
Adil Denizli and Yeşeren Saylan
Department of Chemistry, Hacettepe University, Ankara, Turkey

IOP Publishing, Bristol, UK

© IOP Publishing Ltd 2021

All rights reserved. No part of this publication may be reproduced, stored in a retrieval system or transmitted in any form or by any means, electronic, mechanical, photocopying, recording or otherwise, without the prior permission of the publisher, or as expressly permitted by law or under terms agreed with the appropriate rights organization. Multiple copying is permitted in accordance with the terms of licences issued by the Copyright Licensing Agency, the Copyright Clearance Centre and other reproduction rights organizations.

Permission to make use of IOP Publishing content other than as set out above may be sought at permissions@ioppublishing.org.

Adil Denizli and Yeşeren Saylan have asserted their right to be identified as the authors of this work in accordance with sections 77 and 78 of the Copyright, Designs and Patents Act 1988.

ISBN 978-0-7503-3867-7 (ebook)
ISBN 978-0-7503-3865-3 (print)
ISBN 978-0-7503-3868-4 (myPrint)
ISBN 978-0-7503-3866-0 (mobi)

DOI 10.1088/978-0-7503-3867-7

Version: 20211201

IOP ebooks

British Library Cataloguing-in-Publication Data: A catalogue record for this book is available from the British Library.

Published by IOP Publishing, wholly owned by The Institute of Physics, London

IOP Publishing, Temple Circus, Temple Way, Bristol, BS1 6HG, UK

US Office: IOP Publishing, Inc., 190 North Independence Mall West, Suite 601, Philadelphia, PA 19106, USA

Contents

Preface

We welcome the publication of this book titled *Biosensors for Virus Detection*. In recent years, biosensors have been employed for various applications, from medical diagnosis, environmental monitoring, pharmaceutical analysis, and food quality detection to the defense and security fields. The development and progress of the different types of biosensor require the involvement of chemistry, physics, materials science, nanotechnology, and engineering. A huge body of information on biosensors and applications already exists and continues to enable the creation of more reliable, selective, sensitive, and low-cost sensors for a variety of applications, although the production procedures are complicated and time-consuming.

This book contains eight chapters, which describe biosensors prepared by different methods and used for virus detection applications. In the first chapter, following a discussion of the fundamentals of viruses, the importance of virus detection for medical applications is extensively reviewed. In chapter 2, optical biosensors for virus detection are highlighted. The background and fabrication of optical biosensors and the latest studies of optical biosensors are discussed. Electrochemical biosensors used for virus detection are comprehensively evaluated in chapter 3. Chapter 4 describes the detection of viruses using piezoelectric biosensors documented in recent studies. Chapter 5 covers the use of fluorescence biosensors for medical applications. Chapter 6 summarizes the performance and analytical features of thermal biosensors. In chapter 7, an overview of virus detection using magnetic biosensors can be found. Finally, micromechanical biosensors are examined in detail in chapter 8.

We believe this book provides an overview and highlights some of the recent virus research, including extensively studied hot topics. We would like to extend our deepest gratitude to IOP Publishing and all the contributors who made the creation of this book possible. We hope this book will reach a broad range of readers.

Editors' biographies

Adil Denizli

Professor Dr Adil Denizli was born in 1962 in Ankara. In 1985, he graduated from the Department of Chemical Engineering at Hacettepe University, Ankara, Turkey. He received his Master of Science (M.Sc.) and his Doctor of Philosophy (PhD) degrees in 1986 and in 1992 at Hacettepe University, respectively. He became an Associate Professor of Chemical Technologies in 1994. Dr Denizli has been a Professor in the Department of Chemistry, Hacettepe University since 2000. Dr Denizli has written over 600 articles and 40 book chapters. He has published more than 150 popular science articles. He has four patents. He is one of the most cited scientists in the chemistry and engineering field (over 24000 citations and an h-index of 81). He served as an associated member of the Turkish Academy of Sciences from 1999 to 2007, and as a full member since 2007. Dr Denizli has received 'The Scientific and Technological Research Council of Turkey (TUBITAK) Encouragement Award' in 1998 for his studies of 'Biomedicine, Biotechnology and Environmental Applications of Polymeric Matrices'. He also received the TUBITAK Science Award and the Basic Science Award of Popular Science Magazine in 2006. He has worked as an editor of several national and international books. He was also a chief editor of 15 books marking the 50th year of Hacettepe University. He is an editorial board member of 39 journals. He has supervised 62 MSc and 47 PhD theses, 5 post-doctoral and 5 visiting researchers to date. He has organized various national and international congresses, summer schools, and graduate courses. His main research fields are molecular imprinting technologies, affinity chromatography, biosensors, polymers, and the application of these biosensors and polymers in medicine and biochemistry.

Yeşeren Saylan

Assistant Professor Dr Yeşeren Saylan was born in 1985 in Ordu. She graduated from the Department of Chemistry, Hacettepe University, in 2008. She then received her M.Sc. and PhD degrees from the Division of Biochemistry and the Department of Chemistry, Hacettepe University, in 2011 and 2017, respectively. She worked as a visiting researcher at the School of Medicine, Stanford University, Palo Alto, CA, U.S.A. between 2015 and 2016. She also worked as a visiting researcher at the School of Medicine, Harvard University, Boston, MA, U.S.A. in 2021. Dr Saylan has over 60 publications, including research articles, reviews, and book chapters. She has published more than 20 popular science articles. She is particularly dedicated to developing precise sensing platforms for medical diagnostics, environmental monitoring, and food safety, as well as rationally designing specific molecularly imprinted polymeric materials at the micro- and nano-scales for various applications.

List of contributors

Semra Akgönüllü
Department of Chemistry
Hacettepe University
Ankara
Turkey

Yusuf Aslan
UNAM-National Nanotechnology Research Center
and
Institute of Materials Science and Nanotechnology
Bilkent University
Ankara
Turkey

Suleyman Celik
Sabanci University Nanotechnology Research and Application Center
Sabanci University
Istanbul
Turkey

Umit Celik
School of Civil Aviation
Firat University
Elazig
Turkey

Adil Denizli
Department of Chemistry
Hacettepe University
Ankara
Turkey

Esma Derin
UNAM-National Nanotechnology Research Center
and
Institute of Materials Science and Nanotechnology
Bilkent University
Ankara
Turkey

Özgecan Erdem
UNAM-National Nanotechnology Research Center
Institute of Materials Science and Nanotechnology
Bilkent University
Ankara
Turkey

Cem Esen
Department of Chemistry
Faculty of Arts and Sciences
Aydın Adnan Menderes University
Aydın
Turkey

Muhammad Faisal
HMIS Liaquat University Hospital Hyderabad
Jamshoro
Sindh
Pakistan

Tuğrul Hoşbul
Department of Medical Microbiology
Gulhane Medical Faculty
University of Health Sciences
Ankara
Turkey

Neslihan Idil
Department of Biology
Hacettepe University
Ankara
Turkey

Fatih Inci
UNAM-National Nanotechnology Research Center
and
Institute of Materials Science and Nanotechnology
Bilkent University
Ankara
Turkey

Bo Mattiasson
Division of Biotechnology
Lund University
Lund
Sweden
and
Indienz AB
Billeberga
Sweden

Najma Memon
National Center of Excellence in Analytical Chemistry
University of Sindh
Jamshoro
Pakistan

Tahira Qureshi
National Center of Excellence in Analytical Chemistry
University of Sindh
Jamshoro
Pakistan

Sumera Qureshi
Department of Plant Pathology
Faculty of Crop Protection
Sindh Agriculture University
Tandojam
Sindh
Pakistan

Kutay Sağdıç
UNAM-National Nanotechnology Research Center
and
Institute of Materials Science and Nanotechnology
Bilkent University
Ankara
Turkey

Fatih Şahiner
Department of Medical Microbiology
Division of Medical Virology
University of Health Sciences
Ankara
Turkey

Yeşeren Saylan
Department of Chemistry
Hacettepe University
Ankara
Turkey

Shahnila Shah
National Center of Excellence in Analytical Chemistry
University of Sindh
Jamshoro
Pakistan

Huma Shaikh
National Center of Excellence in Analytical Chemistry
University of Sindh
Jamshoro
Pakistan

Sudhirkumar Shinde
School of Consciousness
Dr Vishwanath Karad Maharashtra Institute of Technology
World Peace University
Kothrud
Pune
India

Saeedreza Zeibi Shirejini
UNAM-National Nanotechnology Research Center
and
Institute of Materials Science and Nanotechnology
Bilkent University
Ankara
Turkey

Abbreviations

3D	Three-dimensional
ab	Antibody
AC	Alternating current
ACE2	Angiotensin converting enzyme 2
AChE	Acetylcholinesterase
ACTES	Triethoxysilylundecanal ethylene glycol acetal
AFM	Atomic force microscope
AFP	Anti-alpha fetoprotein
AgNPs–G	Silver nanoparticles–graphene
AI	Avian influenza
AIDS	Acquired immunodeficiency syndrome
AIV	Avian influenza virus
AlN	Aluminum nitride
ALTES	Triethoxysilylbutylaldehyde
aM	Attomolar
AMR	Anisotropic magnetoresistance
anti-HA Ab	Anti-hemagglutinin antibody
anti-HBc	Total hepatitis B core antibody
anti-HBe	Hepatitis B e-antibody
anti-HBs	Hepatitis B surface antibody
anti-NA Ab	Anti-neuraminidase antibody
APM	Acoustic plate mode
Arbovirus	Arthropod- borne virus
ASF	African swine fever
ASFV	African swine fever virus
Au/MNP-CNTs	Gold/iron-oxide magnetic nanoparticle-decorated carbon nanotubes
Au–SWCNTs	Gold nanoparticles–single walled carbon nanotubes
AuNPs	Gold nanoparticles
$BaTiO_3$	Barium titanate
BAW	Bulk acoustic wave
BSA	Bovine serum albumin
CA	Cell-associated
CARD	Catalyzed reporter deposition
cDNA	Complementary DNA
CDS	Cantilever-driver structure
CEA	Carcinoembryonic antigen
CHIKV	Chikungunya fever virus
CL	Chemiluminescent
CLIA	Chemiluminescence immunoassay
CMV	Cytomegalovirus
CNTs	Carbon nanotubes
CoVs	Coronaviruses
COVID-19	Coronavirus disease 2019
CRISPR	Clustered regularly interspaced short palindromic repeats
CSF	Cerebrospinal fluid
CT	Computed tomography
CV	Cyclic voltammetry

DENV	Dengue virus
DFA	Direct fluorescent antibody
DNA	Deoxyribonucleic acid
DNS-Cl	Dansyl chloride
DPV	Differential pulse voltammetric
dyed-IBV-ab	IBV ab modified by fluorescent dye labeling
EBOV	Ebola virus
ECL	Electrochemiluminescence
EGP	Ebolavirus glycoprotein
EIA	Enzyme immunoassay
EIS	Electrochemical impedance spectroscopy
Env	Envelope glycoprotein
ELIME	Enzyme-linked immuno-magnetic electrochemical assay
ELISA	Enzyme-linked immunosorbent assay
EMPAS	Electromagnetic piezoelectric acoustic sensor
Env	Envelope glycoprotein
EsNCs	Europium sulfide nanocrystals
ET	Enzyme thermistor
EVD	Ebola virus disease
FBAR	Film bulk acoustic resonator
FEA	Finite element analysis
FDA	Food and Drug Administration
FET	Field-effect transistor
FIP	Polyclonal type I feline coronavirus
FISH	Fluorescence *in situ* hybridization
FLISA	Fluorescence-linked immunosorbent assay
fM	Femtomolar
FMDV	Foot-and-mouth disease virus
FMR	Ferromagnetic resonance
FNR	False negative ratio
FPW	Flexural plate wave
FRET	Förster resonance energy transfer
GCEs	Glassy carbon electrodes
GDE	Gold disc electrodes
GF–CTP	Glass fiber–coal tar pitch
GMR	Giant magnetoresistance
GNP–Th	Thionine–gold nanoparticles
GRP	Graphene
HA	Hemagglutinin
HAU	Hemagglutination unit
HAV	Hepatitis A virus
HBeAg	Hepatitis B e-antigen
HBsAb	HBV antibody
HBsAg	Hepatitis B surface antigen
HBV	Hepatitis B virus
HCC	Hepatocellular carcinoma
HCV	Hepatitis C virus
HDV	Hepatitis D
HEV	Hepatitis E
HGV	Hepatitis G

HIV	Human immunodeficiency virus
HPIV	Human parainfluenza viruses
HPV	Human papillomavirus
HRV	Human rhinovirus
HSV	Herpes simplex virus
HTNV	Hantavirus
IBV	Infectious bronchitis virus
ICA	Immuno-chromatographic assay
ICTV	International Committee on Taxonomy of Viruses
IDT	Interdigital transducer
IFA	Indirect fluorescent antibody
Ig	Immunoglobulin
IgE	Immunoglobulin E
IgG	Immunoglobulin G
IgM	Immunoglobulin M
JEV	Japanese encephalitis virus
LA	Linear array
LAMP	Loop-mediated isothermal amplification
LBA	Line blot assay
LED	Light-emitting diode
LFA	Lateral-flow immunoassay
LIA	Line immunoassay
LIPA	Line-probe assay
$LiTaO_3$	Lithium tantalate
LOD	Limit of detection
LOQ	Limit of quantification
L-SAW	Lamb surface acoustic wave
LSPR	Localized surface plasmon resonance
mAb	Monoclonal antibody
mABs	Monoclonal antibodies
MAC-ELISA	IgM-capture enzyme-linked immunosorbent assay
MB	Magnetic bead
MEF	Metal-enhanced fluorescence
MEMS	Micro-electromechanical systems
MERS	Middle East respiratory syndrome
MERS-CoV	Middle East respiratory syndrome coronavirus
MFLI	Lock-in-amplifier
MHDA	16-mercaptohexadecanoic acid
MIL-101	Silicon-modified metal–organic framework
MIP	Molecularly imprinted polymer
MLPA	Multiplex ligation-dependent probe amplification
MMB	Magnetic modulation biosensing
MnMg-NFPs	Nanoferrite particles
MNPs	Magnetic nanoparticles
MOF	Metal–organic framework
MOSFET	Metal–oxide–semiconductor field-effect transistor
MPs	Magnetic particles
MR	Magnetoresistance
MRI	Magnetic resonance imaging
MUA	Mercaptoundecanoicacid

MUT	Micromachined ultrasonic transducer
MWCNTs	Multiwalled carbon nanotubes
NA	Neuraminidase
NAATs	Nucleic acid amplification tests
NASBA	Nucleic acid sequence-based amplification
NAT	Nucleic acid test
NEMS	Nanoelectromechanical Systems
NH2-CC-MNPs	Amino-functional carbon-coated magnetic nanoparticles
NHS	n-hydroxysuccinimide
nM	Nanomolar
NMR	Nuclear magnetic resonance
NoVs	Noroviruses
NP	Nanoparticle
OBD	Optical beam deflection
OLA	Oligomer ligation assay
PAA	Polyacrylamide
PBS	Phosphate-buffered saline
$PbTiO_3$	Lead titanate
PC	Pyrene-1-carboxaldehyde
PCR	Polymerase chain reaction
PDMS	Polydimethylsiloxane
PEF	Plasmon-enhanced fluorescence
PEG	Polyethylene glycol
PEI-Glu	Polyethyleneimine glutaraldehyde
PET	Positron emission tomography
p-FLISA	Plasmon-enhanced fluorescence-linked immunosorbent assay
PFU	Plaque-forming unit
PLLA	Polylactic acid
pM	Picomolar
PMGRP	Plasma magnetic graphene
PMMA	Polymethylmethacrylate
PNA	Peptide nucleic acid
POC	Point of care
PRNT	Plaque reduction neutralization test
PVDF	Polyvinylidene fluoride
PZT	Lead zirconate titanate
QCM	Quartz crystal microbalance
QDs	Quantum dots
qRT-PCR	Quantitative reverse-transcription PCR
RCA	Rolling circle amplification
rGO	Reduced graphene oxide
RIA	Radio immunoassay
RIBA	Recombinant immunoblot assay
RIDT	Rapid influenza diagnostic tests
RN	Red nanoparticle
RNA	Ribonucleic acid
R-SAW	Rayleigh surface acoustic wave
RSV	Respiratory syncytial virus
RT-PCR	Reverse-transcription polymerase chain reaction
Rt-PCR	Real-time PCR

SAM	Self-assembly monolayer
SARS	Severe acute respiratory syndrome
SARS-CoV	Severe acute respiratory syndrome coronavirus
SARS-CoV-2	Severe acute respiratory syndrome-related coronavirus 2
SARS-CoV-2019	Severe acute respiratory syndrome coronavirus 2019
SAW	Surface acoustic wave
SDA	Strand-displacement amplification
SEF	Surface enhanced fluorescence
SEM	Scanning electron microscopy
SERS	Surface-enhanced Raman scattering
SH-APM	Shear horizontal acoustic plate mode
SH-SAW	Shear horizontal surface acoustic wave
SiO_2	Quartz
SL-PAA	SL-polyacrylamide
SNR	Signal-to-noise ratio
SPECT	Single photon emission computed tomography
SPEs	Screen-printed electrodes
SPGEs	Screen-printed gold electrodes
SPR	Surface plasmon resonance
SQUIDs	Superconducting quantum interference devices
ssDNA or ssRNA	Single-stranded oligo-nucleic acids
STW	Shear transverse wave
SWCNT	Single-walled carbon nanotube
SWV	Squarewave voltammetryTAT
TAT	Transactivator of transcription
TBA15-1	Thrombin DNA aptamer
TCID	Tissue culture infective dose
$TCID_{50/mL}$	Median tissue culture infectious dose
TEOS	Tetraethyl orthosilicate
TEP	Thermal enzyme probes
Tg	Glass-transition temperature
TMA	Transcription mediated amplification
TMR	Tunneling magnetoresistance
TO	Thiazole orange
TSM	Thickness shear mode
Vds	Drain–source voltage
VP73	Virus protein 73
VZV	Varicella-zoster virus
WHO	World Health Organization
ymol	Yoctomole
zM	Zeptomolar
zmol	Zeptomole
ZnO	Zinc oxide

IOP Publishing

Adil Denizli and Yeşeren Saylan

Chapter 1

Viruses and the importance of virus detection

Tuğrul Hoşbul and Fatih Şahiner

While discoveries about viruses, which carry the largest genetic information code in nature, have continued unabated for more than a century, viruses are constantly affecting the global biological balance. The main characteristics that make viruses important in this balance include epidemics and mass deaths, the risks of being used as biological weapons, and the continued lack of an effective vaccine or specific treatment for many viruses. However, viruses also offer incredible opportunities for the scientific world, due to biotechnological innovations that have gained momentum in recent years. The detection of viruses is of urgent importance for the early detection of pandemics and epidemics. In this chapter, we discuss the classification of viruses, the effects of viruses on biological balance of our planet, the importance of detecting viruses that have the potential to be used as biological weapons or that carry epidemic or pandemic risks, and the characteristics of such viruses.

1.1 Introduction

Viruses are non-living micro-organisms that are only able to replicate in living cells [1, 2]. However, a given virus cannot replicate or cause disease in all living organisms, because the replication of viruses is restricted by host and tissue tropism [3]. Different types of virus infect different cellular organisms [2, 4]. Some viruses are able to infect fungi, plants, archaea, and bacteria, but these viruses do not cause infection or disease in humans [2, 3]. According to Current Taxonomy Release 2019, produced by the International Committee on Taxonomy of Viruses (ICTV), more than 6500 viruses have been identified at the species level [5]. More than 300 of these viruses are associated with human infection [3], but not all viral infections result in disease. Viruses that cause disease are defined as pathogenic species, and the number of virus species that causes disease in humans is over 200 [4, 6]. Pathogenic viruses that spread among humans and animals can cause serious mortality and morbidity, as well as significant economic burdens [7].

doi:10.1088/978-0-7503-3867-7ch1 © IOP Publishing Ltd 2021

In the last 30 years, there have been significant changes in the epidemiology of viruses associated with human infections. While many viruses (such as the West Nile virus) have spread to regions where they had never been detected before, others have caused global concern after a long period of silence, such as the Zika virus [8, 9]. Apart from this, new viruses have emerged and caused regional and global epidemics, such as the Middle East respiratory syndrome-related coronavirus (MERS-CoV) and severe acute respiratory syndrome-related coronavirus 2 (SARS-CoV-2) [10]. Moreover, due to developments in molecular biology and genetics, it is also important to monitor the natural viruses and modified viruses that can be manipulated more easily and which have the potential to be used as biological warfare agents [11, 12].

Viruses are mainly transmitted to humans through the respiratory tract, food and drink, skin, and blood [3]. Some viruses are transmitted to humans by insects, such as mosquitoes and ticks, and mammals, such as cats, dogs, mice, bats, and birds [3, 7]. Since viruses are host-dependent micro-organisms, their distribution in geographic regions is directly related to their host distribution [3]. Humans are the only hosts of some viruses in nature; such viruses, which do not have an animal reservoir, continue their existence by spreading among people [3, 4]. Cultural habits, nutritional and sanitation conditions, the frequency of risky behaviors, drug addiction, the proportion of the immunodeficient population in a society, access to health services, vaccination programs, and other health policies are some of the social characteristics that define the boundaries of the epidemiology of these viruses. On the other hand, in the distribution of animal viruses associated with human infections (zoonotic viruses), the possibility of contact between humans and animals (for example, the intersection of the habitats of humans and reservoir animals) plays an important role in viral epidemiology [7]. The effects of climate change on the circulation of migratory birds and arthropods such as mosquitoes, human movements that have accelerated with the increase in transportation facilities, and forced migration are other factors that determine the distribution boundaries of viral diversity in human populations [7].

1.2 Importance of virus detection

Typical (pathognomonic) symptoms with diagnostic value can be seen in some viral infections. However, in most viral infections, the clinical findings are similar to those for multiple viruses, and sometimes infections can take place without any symptoms (asymptomatic) [13, 14]. Therefore, precise diagnoses of viral infections are carried out using laboratory tests (figure 1.1). Viral infections can be diagnosed directly by methods such as the visualization of viruses, the detection of viral antigens, the isolation of a virus by replicating it in a cell culture medium, and by the observation of viral nucleic acids. Viral infections can also be identified indirectly by different methods, such as the detection of virus-specific antibodies, the examination of changes (such as inclusion bodies) in infected cells using light microscopy, and the indirect examination of growths in cell cultures [15]. Laboratory tests are also used for different clinical purposes, such as monitoring the patient in some infections,

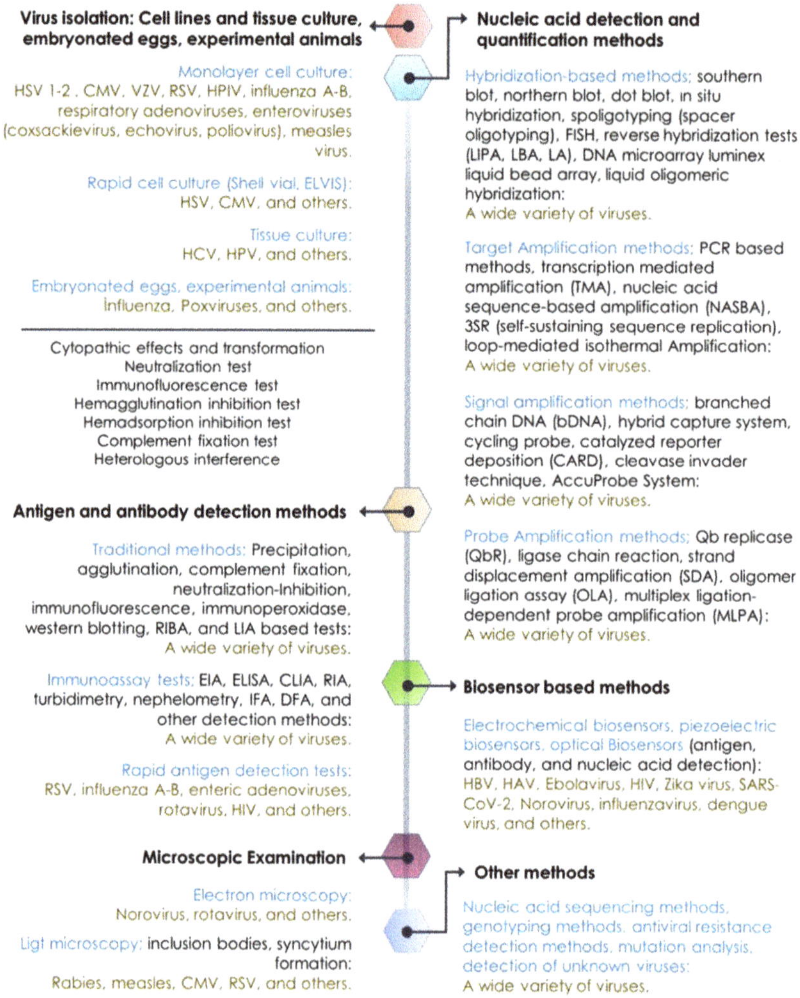

Figure 1.1. Overview of the methods used to detect and examine viruses. Abbreviations: HSV 1–2 (herpes simplex virus), CMV (cytomegalovirus), VZV (varicella-zoster virus), RSV (respiratory syncytial virus), HPIV (human parainfluenza viruses), HAV (hepatitis A virus), HBV (hepatitis B virus), HCV (hepatitis C virus), HPV (human papillomavirus), HIV (human immunodeficiency virus), RIBA (recombinant immunoblot assay), LIA (line immunoassay), EIA (enzyme immunoassay), ELISA (enzyme-linked immunosorbent assay), CLIA (chemiluminescence immunoassay), RIA (radioimmunoassay), IFA (indirect fluorescent antibody), DFA (direct fluorescent antibody), FISH (fluorescence *in situ* hybridization), LIPA (line-probe assays), LBA (line blot assay), LA (linear array).

investigating the presence of antiviral resistance, and distinguishing between primary and reactivated infections. The detection of viruses is also critically important in epidemic outbreak surveillance and other epidemiological and environmental analyses, phylogenetic studies, the screening of blood and blood products, veterinary studies, and tests for biosecurity breaches (due to the risk that a virus may be used as a biological weapon) [12, 16].

Nowadays, new molecular techniques that allow the detection of unknown microorganisms, specific cell lines, and chimeric cultures, animal models that increase the probability of isolation of viruses, and metagenomic-metaviromic studies have increased the number of known viruses and made it possible to monitor genomic variations and epidemiological changes in these viruses [16, 17]. The fast detection of pathogens, using methods that offer specific, sensitive, fast, and reproducible results, plays an important role in preventing the spread of viral infections and decreasing the pathologies (mortality and morbidity) associated with infection [18]. Rapid identification of the presence of a virus in contaminated food or water, surfaces, or patient samples is a prerequisite for effectively preventing viral outbreaks, epidemics, and bioterrorism [12, 18]. As we have experienced in the SARS-CoV-2 pandemic, it is far from realistic for developed countries to isolate themselves from the rest of the world due to intense and rapid human movement. For these reasons, the development of easy-to-use, low-cost detection methods that require fewer complex devices and equipment in epidemic management and isolation-based public health policies is critical in the management of global epidemics [19].

1.2.1 Coronavirus

The name coronavirus is derived from the Latin word 'corona,' meaning halo, which expresses the similarity of the characteristic appearance of the surface protrusions of the virus to the halo (solar corona) around the Sun [20]. There are 46 virus species in five genera in the *Coronaviridae* family [5]. Six of these species were known to be associated with human infection. In the last days of 2019, a new coronavirus identified as a pneumonia agent, which first appeared in Wuhan, China, spread rapidly among people, and the number of coronavirus species associated with human infection was updated to seven [21]. Among these viruses, HCoV-229E (mid-1960), HCoV-OC43 (mid-1960), NL63 (in 2004), and HCoV-HKU1 (in 2005) are known to be seasonal coronaviruses that are common worldwide and which cause respiratory infections that are generally associated with mild symptoms in humans [22]. Unlike seasonal coronaviruses, the newly emerging zoonotic coronaviruses cause severe respiratory infection and death (with a mortality rate of 1%–35%) [23, 24]. Among these viruses, SARS-CoV was first isolated in China in 2003 and caused 8096 cases and 774 (9.6%) deaths in 27 different countries within a year [23]. SARS-CoV disappeared after spreading among humans for one year, and no new human infections were reported worldwide after 2004. Another virus, MERS-CoV, was first identified in Saudi Arabia in 2012 and infected 2449 people in more than twenty countries between 2012 and 2019, causing 845 (35%) deaths [24]. The SARS-CoV and MERS-CoV viruses attracted attention by causing epidemics, especially in hospital settings [25, 26]. These two viruses did not cause pandemics, as they were limited to certain regions, but both viruses were examined in detail due to their pandemic potential, and a lot of research has been done on these two viruses. The experience gained in these studies has been of critical importance in fighting the pandemic coronavirus SARS-CoV-2, which causes the disease known as COVID-19 (coronavirus disease 2019) [21]. SARS-CoV-2, which has spread more easily among humans, reached all parts of the world in a short time, and is now described as one of the most serious epidemics of the last hundred years, with close to 120 million confirmed cases and 2.6 million deaths [19, 27].

The major transmission route of coronaviruses in humans, which is similar to those of other respiratory viruses, is through the inhalation of droplets containing infective viral particles produced when an infected person speaks, coughs, or sneezes. Although it is not a predominant transmission mode, touching the mouth, nose, or eyes (autoinoculation) after contact with virus-contaminated surfaces or fomites is also considered to be one of the possible transmission routes [28]. The most effective way to prevent coronavirus infections, which exhibit similar transmission characteristics to those of other respiratory infections, includes standard measures taken to prevent droplet and aerosol transmission. Inactivated full-virus vaccines, messenger RNA (ribonucleic acid) vaccines, and vector-based vaccines developed within about a year of the onset of the SARS-CoV-2 epidemic have been administered to millions of people worldwide [29]. Even if these vaccines do not prevent the transmission of infection, they are expected to prevent serious illnesses requiring hospitalization and SARS-CoV-2 related deaths and slow the spread of the epidemic. The protection period and protective effectiveness of vaccines in different populations are expected to be revealed in upcoming studies.

Laboratory diagnosis of coronavirus infections is based on tests that aim to detect viral RNA using a PCR (polymerase chain reaction) and viral antigens or specific antibodies against the virus [30]. Virus isolation is time-consuming; it is not used in routine diagnosis and also requires a biosafety level 3 (BSL-3) laboratory using BSL-3 practices [31]. During the COVID-19 pandemic, it has been of great importance for the rapid and correct diagnosis of cases, for the detection and isolation of people with asymptomatic infections who can spread the infection, and to keep the pandemic under control. For this purpose, many biosensor-based methods have also been developed [32].

1.2.2 Human immunodeficiency virus

There are ten different viruses in the *Lentivirus* genus of the *Retroviridae* family, and two of these viruses, human immunodeficiency virus 1 (HIV-1) and HIV-2, are associated with human infection [5]. The disease known as acquired immunodeficiency syndrome (AIDS) was first reported in 1981 in the United States of America (USA) in a group of homosexual men with *Pneumocystis jirovecii* pneumonia and Kaposi's sarcoma [33, 34]. In 1982, some researchers from the CDC (Centers for Disease Control and Prevention) described similar cases in injected-drug users, sex workers, and people using blood and blood products [35]. As a result of subsequent studies, the HIV-1 and HIV-2 viruses were found to be infectious agents for AIDS in 1983 and 1986, respectively [36]. HIV-1 accounts for the vast majority of human lentivirus-originated disease worldwide. HIV-2, whose genetic composition and disease symptoms are very similar to HIV-1, is endemic in West Africa and is seen in the USA, albeit at low rates. Within days or weeks following exposure to HIV, a disease picture that is defined as primary HIV infection (acute viral syndrome) and which usually presents acute flu-like symptoms occurs in most infected individuals [37]. AIDS patients typically show signs of a selective decrease in the CD4+ helper-T lymphocytes in their peripheral blood samples and immune system failure, which is due to this decrease [38, 39]. After acute infection, due to the establishment of a balance between viral replication and host immune response, a latency period begins, during which clinical symptoms are not seen for many years in most infected individuals [37, 38]. However, the phrase 'latency period' can be misleading, given the very high rate of virus replication and the high daily production–destruction rate of CD4+ T-cells. Because this process causes a gradual decrease in CD4+ T-cells, albeit slowly, when the CD4+ T-cell ratio falls below 200 μL^{-1}, many AIDS-defining diseases, including various opportunistic infections and some cancers, are seen, and the process results in death in untreated cases [37, 40].

According to the Joint United Nations Program on HIV/AIDS (UNAIDS) 2020 report, the number of people living with HIV worldwide is estimated to be approximately 38 (31.6–44.5) million [41]. It is estimated that a total of 32.7 million (24.8–42.2) people have died from AIDS-related diseases from the beginning of the epidemic to the end of 2019. For 2019, the number of newly infected people is estimated at 1.7 million and the number of HIV-related deaths at 690 000. Although an effective treatment exists today, these high mortality rates seem to be related to the inability to control the infection in regions where the majority of infected individuals have limited access to treatment [41]. All these data highlight that HIV is one of the most important viral infections that has affected humanity in the last 40 years.

HIV infection is transmitted mainly by unprotected sexual intercourse with an infected partner, the injection or transfusion of contaminated blood or blood products, infertility treatment practices, skin grafting, organ transplantation, the sharing of non-sterile injection equipment previously used by an infected person, and mother-to-baby transmission (during pregnancy, birth, and breastfeeding) [34, 42]. A preventive vaccine has not yet been developed for HIV infection. To reduce the effects of infection and to control or eliminate HIV-1 reservoirs, there is increasing interest in the use of agents that activate the latent virus, therapeutic vaccines, widely neutralizing monoclonal antibodies, gene therapy, and a variety of other pharmacological and immunological approaches [43].

Many different tests can be used to diagnose acute HIV-1 infection. Laboratory diagnosis of HIV infection is based on the detection of virus and virus components including viral deoxyribonucleic acid (DNA), RNA, and antigens, as well as antibodies against the viral antigens in the blood [38, 44]. The diagnosis of acute infection is based on the detection of HIV-1 replication by molecular methods in the absence of HIV-1 antibodies (as they are not yet present in the early stages of infection). These highly sensitive tests (branched-chain DNA, PCR, Gen-Probe) work on the principle of detecting the presence of HIV-1 RNA in plasma [45]. The diagnosis of acute HVI-1 infection by molecular tests should be confirmed within the following weeks by a positive HIV-1 antibody test (seroconversion). Another diagnostic approach is the detection of viral antigens. Tests based on p24 antigen detection have been developed for this purpose [37]. Today, combo tests that work with the quantitative chemiluminescence method that detect both antibodies against HIV-1, -2, and p24 antigen are used [46]. New strategies are also being studied to assess the effectiveness of new therapies, for example, by detecting changes in the number of HIV-1-infected cells and determining the transcriptional activity of these cells by estimating their cell-associated (CA) HIV-1 DNA and CA HIV-1 RNA levels [43].

1.2.3 Hepatitis

During many viral infections, liver tissue can be directly or indirectly affected, and hepatitis may develop. However, the definition of hepatitis viruses is used to refer to viruses that show tropism directly to the liver: the hepatitis A virus (*Hepatovirus* A), the hepatitis B virus, the hepatitis C virus (*Hepacivirus* C), the hepatitis D virus (of the hepatitis genus *Deltavirus*), and the hepatitis E virus (*Orthohepevirus* A) [5, 47]. Each of these viruses is classified as belonging to a different family, and they are RNA viruses except for the hepatitis B virus, whose genome structure is DNA [5]. Among these viruses, hepatitis B and C viruses can cause chronic infection, liver disease, cirrhosis, and even hepatocellular carcinoma (HCC) in the long term [48, 49]. Chronic hepatitis B and C infections are the most important causes of HCC, and these two viruses constitute the majority of cases (80%) worldwide [50].

Although hepatitis A, B, C, D, and E viruses all cause liver disease, they differ in important ways, such as their transmission patterns, the severity of the disease, geographic distribution, and prevention [47]. While hepatitis A and hepatitis E viruses are common in countries with poor sanitation conditions in the world, hepatitis B and C are seen all over the world, being more common in some parts of Africa, Asia, and South America [51]. It is estimated that 325 million people worldwide are living with hepatitis B and/or C, and that most are without access to diagnosis and treatment. Hepatitis A and E viruses are mainly transmitted by the fecal–oral route, while the hepatitis B, C, and D viruses are mainly transmitted by the parenteral route [49]. For the hepatitis B virus, perinatal transmission and horizontal transmission (exposure to infected blood) are important transmission routes, while needlestick injury, exposure to infected blood and body fluids, and sexual transmission are other transmission routes [47]. The hepatitis C virus, on the other hand, spreads via blood and injected-drug use through the sharing of injection equipment [28, 47]. While there is a preventive vaccine for hepatitis A, B, and E (the hepatitis E vaccine is used in China), a preventive vaccine has not yet been developed for the hepatitis C virus [48, 51]. Although powerful treatment protocols have been developed and are in use for the hepatitis B and C viruses, which are associated with chronic diseases, access to treatment is easier in developed countries, but a large proportion of infected people is deprived of access to treatment [49].

Although clinical signs and nonspecific laboratory data are of some benefit in the diagnosis of hepatitis viruses, high-sensitivity tests for the detection and quantitation of viral nucleic acids and the analysis of the presence of immunoglobulin M (IgM) and immunoglobulin G (IgG) against viral antigens are used in the specific laboratory diagnoses of these viruses [17, 52]. In developed countries, blood and blood products are routinely screened for the hepatitis B and C viruses to minimize the transmission of hepatitis viruses through blood [47]. Quantitative molecular tests have also been developed for the hepatitis B and C viruses, and these tests have become critical tools for clinicians in the monitoring of antiviral therapies [17]. In addition, genotype determination and antiviral resistance monitoring are important approaches used in the management of diseases caused by these viruses [17, 52].

1.2.4 Ebola

There are six genera and 11 species in the *Filoviridae* family; one species in the *Marburgvirus* genus in this family (*Marburg marburgvirus*) and four species in the *Ebolavirus* genus are associated with human infection [5]. The *Ebolavirus* species isolated on the African continent, namely, the Bundibugyo Ebola virus, the Sudan Ebola virus, and the Zaire Ebola virus cause outbreaks with high mortality in humans, but Taï Forest Ebola virus infections are rare [53]. Although there are serological data on human infections for Reston Ebola virus, another species first isolated in macaques in the Philippines, human–human transmission has not been demonstrated [54]. Ebola virus disease (EVD) is an acute severe illness with a high mortality rate caused by pathogenic Ebola virus strains. The mortality rate has ranged from 25% to 90% in various outbreaks of the disease over 40 years [53].

The Ebola virus was first identified in 1976 in Zaire (now the Democratic Republic of the Congo (DRC)) and South Sudan during two concurrent outbreaks. In these epidemics, a total of 602 people were infected with Ebola virus and the mortality rate was found to be 71.59% [53]. On March 23, 2014, the West African epidemic broke out, which represented the first global threat posed by Ebola virus. A few months later, the World Health Organization (WHO) stated that this epidemic was an international 'public health emergency.' In the biggest Ebola virus outbreak, 28603 EVD cases were reported with a mortality rate of 39.5% [53, 55, 56]. Although Ebola virus infections are only common in West Africa, the virus is classified as a category A pathogen that could be abused as a bioterrorism agent due to its potential to spread globally [57].

Ebola virus infections can be transmitted in the following ways: via the contact of infected organisms with tissue, blood, and other body fluids, via contaminated materials or medical devices, and during unhygienic practices (such as cutting up a chimpanzee for food or traditional burial practices) [53, 58]. The Ebola virus is classified at level 4 in the biosafety risk group; when working with it in the laboratory, facilities and equipment that provide the highest level of biosecurity are required [53, 55].

The laboratory diagnosis of Ebola virus can be accomplished by examining blood, body fluids, and other clinical specimens from animals and humans [53]. Enzyme-linked immunosorbent (ELISA) tests, antigen detection tests, serum neutralization tests, and reverse-transcription PCR can be used to examine these samples [55, 57, 59]. IgM and IgG antibodies can be measured using immunofluorescence antibody (IFA), or immunoassay-based tests [53, 58, 59].

1.2.5 Zika

Zika virus is one of the 53 different species of the *Flaviviridae* family in the *Flavivirus* genus [5]. It was first discovered in the Zika forest (Kampala, Uganda) in 1947 [9]. The majority of Zika virus infections have a subclinical course, but some patients may experience clinical symptoms similar to those of other arboviral infections [60]. Eighty percent of infected people are asymptomatic [61]. The disease symptoms are usually mild and are characterized by the acute onset of mild fever, maculopapular rash, arthralgia, and non-purulent conjunctivitis [61, 62]. Until 2015, no deaths were attributed to Zika virus infections. Three deaths attributed to the Zika virus were first reported in Brazil, including the death of a newborn with microcephaly, on 28 November 2015 [63]. Also, various neurological (Guillain Barré syndrome, meningoencephalitis) and autoimmune (thrombocytopenic purpura, leukopenia) complications have been reported in Zika-virus-infected people [64]. While the Zika virus has attracted great attention in recent years due to the epidemics it causes and its possible relationship with serious autoimmune–neurological diseases, it was identified as an urgent public health problem of international importance by the WHO in February 2016 [9, 64–66].

Identified in 1947 by researchers conducting yellow fever surveillance in Uganda, until recently, the Zika virus was known as a infectious agent limited to tropical regions in Africa and Asia [60, 67, 68]. Until the first major epidemic in Micronesia's Yap Island in 2007, very few cases of Zika virus were reported. Zika virus infection has become a continental and global health problem, with outbreaks in South and Central America in May and November 2015. However, when virological studies, seroprevalence studies, diagnosis of sporadic cases, and epidemic data were evaluated together, it was shown that the Zika virus has spread over a wide area including Asia, Africa, the Pacific Islands, South and Central American countries, and the Caribbean [9].

The Zika virus is primarily transmitted by the bite of infected mosquitoes of the *Aedes* genus [61]. Mosquitoes become infected by feeding on a viremic human and the virus is transmitted to other people through the bites of infected mosquitoes. In addition, it can be transmitted from a pregnant mother to her baby during pregnancy or on the days close to birth [66, 69]. Transmission through infected blood or sexual contact is also possible; to date, a small number of cases have been reported regarding the possible transmission of the virus via blood transfusion and the possible spread of the virus through sexual contact [70].

There are some difficulties in the diagnosis of Zika virus infections. Clinical diagnosis is difficult because the clinical presentation is similar to those of other arbovirus infections (such as chikungunya and dengue). Clinicians should undertake a differential diagnosis with confirmatory tests of people who live in the area where Zika virus infections are seen or who have a travel history [71]. Another difficulty is the presence of heterologous antibodies and complex disease presentations with different degrees of clinical pathology in co-infections caused by the Zika virus and similar viruses [64, 71, 72]. This situation makes it more difficult to diagnose Zika virus infections, which already require additional tests in the differential diagnosis, since they may cross-react with other flaviviruses in serological tests. The main methods used in the diagnosis of Zika virus infections are: investigating the presence of viral RNA using reverse-transcription PCR in various clinical specimens including serum, cerebrospinal fluid (CSF), amniotic fluid, urine, semen, saliva, placental, and fetal tissues of patients with any clinical symptoms, along with sequence analysis, genotyping studies, investigating viral antigens in tissue samples by immunohistochemical methods, detecting the presence of viral IgM and IgG in patient serum or CSF samples using various serological tests, differential diagnosis, and confirmation using the plaque reduction neutralization test (PRNT) [60, 62, 67, 69, 71, 73].

1.2.6 Norovirus

Norwalk virus, the only species in the Norovirus genus, is a non-enveloped, positive-sense RNA virus classified as being in the *Caliciviridae* family [5]. Another species associated with human infection in the *Caliciviridae* family is the Sapporo virus, the only species in the *Sapovirus* genus [74]. Both viruses generally induce self-limiting gastroenteritis. Long-term illness is seen in immunocompromised individuals. Vomiting is a consistent and remarkable symptom; other symptoms may include nausea, diarrhea, abdominal cramping, fever, and malaise. Norwalk virus causes a brief but unpleasant illness, also known as stomach flu, viral gastroenteritis, and winter vomiting [74–77].

Norwalk virus was first described in 1968 at a school in Norwalk, Ohio [76, 77]. This virus is the leading cause of norovirus gastroenteritis epidemics worldwide [78]. New norovirus strains appear every two to four years; these new strains frequently, but not always, lead to an increase in worldwide epidemics [79]. Noroviruses are usually transmitted by contaminated food or water, but can also be spread by person-to-person contact [76, 77]. These viruses are very contagious because even a few particles can cause disease, and infected individuals become highly contagious. Transmission occurs via fecal–oral exposure to contaminated food and water or infectious aerosols produced by vomiting [80]. As a result of this high infectivity and efficient transmission, newly emerging strains of norovirus could lead to global epidemics [78]. It inevitably causes epidemics in crowded areas such as hotels, dormitories, hospitals, and cruise ships [80].

Diagnostic methods for norovirus focus on the detection of viral RNA (mainly using PCR-based tests) or viral antigens, since noroviruses are difficult to propagate in cell cultures, and antibody tests are not significant in the diagnosis of acute infection [74]. Conventional reverse-transcription PCR followed by sequence analysis is used for norovirus genotyping [79]. Radioimmunoassay (RIA) and ELISA tests for antibody detection can be used in epidemiological studies. With immunoelectron microscopy, diagnosis is also possible by directly observing the virus, but it is not used in routine practice [74].

1.2.7 Influenza

Viruses in the *Orthomyxoviridae* family are enveloped and have single-stranded segmented RNA genomes [81]. This family includes viruses that can be transmitted by respiration, direct contact, water, or ticks [82]. Orthomyxoviruses can cause infection in humans, birds, many mammal species, and fish. The most medically important viruses in this family are the influenza virus species. These viruses (influenza viruses A, B, and C) are associated with respiratory tract infections that can lead to serious clinical consequences in humans [81, 82]. Influenza viruses can lead to seasonal epidemics and sometimes pandemics (influenza A). Throughout history, influenza A pandemics have been blamed for the deaths of tens of millions of people. Birds are the main reservoir of influenza A viruses, and humans and many animal species (such as pigs, horses, seals, whales, and mink) are affected by the virus [83]. Very virulent or pandemic reassortant strains may arise when influenza A viruses that cause infections in different species cross the species barrier [4, 82].

Seasonal influenza is an acute respiratory infection caused by influenza viruses circulating around the world [83]. There are four types of seasonal influenza virus: influenza viruses A, B, C, and D [84]. Influenza A and B viruses circulate around the world and are associated with seasonal epidemic diseases. Some influenza A viruses can cause a pandemic, usually at intervals of 20–30 years. An influenza virus pandemic occurs when a new strain circulates among humans, since most people are not immune to this new virus. This situation allows the new strain to spread easily among humans [83]. Seasonal influenza viruses can contribute to the emergence of a pandemic virus, for example, pandemic A H1N1 in 2009. When a pandemic virus spreads globally and becomes established, it may become a seasonal virus and can remain in circulation for many years [85]. Influenza virus B, which was first isolated in 1940, is constantly circulating in humans and causes recurrent epidemics of respiratory disease. Antigenic change (antigenic drift) occurs more slowly in influenza virus B than in influenza virus A [86]. Influenza virus B variants have only been detected in humans (no known animal reservoirs are widely circulated), almost exclusively infect humans, and are therefore not associated with pandemic risk [82, 87, 88]. Member viruses of the influenza virus genus C naturally infect humans [82]. Influenza C virus is less common and usually causes mild sporadic respiratory illness, so it is of low public health importance and not associated with pandemics [83]. Influenza D viruses mainly affect cattle and are not known to cause infection or disease in humans [85]. Illnesses range from mild to severe (and even death). Hospitalization and death are mainly seen among high-risk groups. Worldwide, seasonal influenza outbreaks are estimated to result in approximately 3 to 5 million serious illness cases and approximately 290 000 to 650 000 deaths annually [85, 89].

The natural transmission of influenza viruses occurs by aerosol (most human and non-aquatic hosts) or by water (waterfowl), but direct contact may also be significant [83]. Influenza A spreads via the respiratory system and animal contact among mammals, but is transmitted via the fecal–oral route by contaminated water among birds [82]. The influenza virus is transmitted from person to person by contact with airborne droplets or contaminated hands or surfaces. During the influenza season, influenza viruses circulate in the community; the best way to avoid influenza is to be vaccinated every year. Preventive vaccines are available for influenza A and B and are particularly recommended for people at risk of severe illness [83]. The annual seasonal influenza vaccine is the best way to reduce the risk of seasonal influenza and prevent transmission to other people [89].

Cell culture, direct fluorescence antibody (DFA), rapid diagnostic tests, PCR, and antibody-based serological tests are used for diagnosis [83]. Rapid influenza diagnostic tests (RIDTs) that work by detecting viral antigens can be applied in 10–15 min, but the sensitivity of these tests is low. Therefore, even if a rapid test result is negative, a patient may still have influenza. Rapid molecular analyses are more accurate than RIDTs [89].

1.2.8 Other viruses

Viruses are important to humanity for a variety of reasons. Smallpox virus, Nipah virus, Lassa fever virus, Rift Valley fever virus, Crimean Congo hemorrhagic fever virus, Ebola virus, and Marburgvirus are important as biological warfare agents [11, 12], while some viruses are critical, as they have pandemic potential (influenza viruses, coronaviruses) [10]. Congenital infections (such as cytomegalovirus, rubella, and parvovirus), childhood diarrheal diseases (rotavirus), recurrent or reactivated infections in cases of immunodeficiency (human cytomegalovirus, JC virus, BK virus, and Epstein-Barr virus), chronic diseases (the hepatitis B and C viruses, HIV), and viruses that lead to economic damage to agriculture and veterinary medicine (rabies virus, foot and mouth disease virus, influenza virus, and Newcastle disease virus) are further examples.

Viruses are classified into two main groups, DNA and RNA viruses, based on their genetic structures [1]. RNA viruses usually cause transient infections and are cleared by the immune system, however, there are exceptions to this. Hepatitis C virus is an RNA virus that can be completely cleared from the body, but if left untreated, it can become permanent, because it can escape from the control of the immune system [48]. Another exceptional group of RNA viruses can cause late or prolonged central nervous system infections (slow infections) [40]. These viruses include the measles virus, the rubella virus, Nipah virus, and the rabies virus, which can escape from the immune system for a period of 1–15 years [8]. The third and most important exception is HIV, the causative agent of AIDS. Although HIV is an RNA virus, the genetic code of the virus transforms into a DNA structure in the human body, reaches the cell nucleus, and becomes permanently integrated into the chromosomal structure (provirus), and when the host cells divide and multiply, this virus continues to reproduce with the infected host cells [1, 38]. In cases of immunosuppression or immunodeficiency, chronic or prolonged infection caused by RNA viruses may also occur; hepatitis E virus is a good example of such RNA viruses [47].

Human and animal DNA viruses are generally species specific [3]. DNA viruses replicate in the cell nucleus, but exceptionally, poxviruses replicate in the cell cytoplasm. In general, DNA viruses cannot be transmitted from animals to humans; only some poxvirus species are known to cause zoonotic infections [90]. In contrast to viral RNA, the human immune system does not react strongly to the DNA of these viruses; and most of these viruses cause permanent latent infections in the host, remain silent in cells for life, and hide from the immune system [40]. However, these viruses can cause life-threatening illnesses in situations in which the immune system is suppressed, such as human cytomegalovirus, Epstein-Barr virus, JC virus, and BK virus. It is also known that some DNA viruses including the hepatitis B virus, the human papillomavirus, the Epstein-Barr virus, and some polyomavirus species are associated with specific cancer types following chronic persistent infections [40, 47, 91].

1.3 Conclusions and future perspectives

The early and rapid identification of viruses that are emerging, have epidemic potential, cause permanent damage, or pose a biosecurity risk is critical. These risks are increasing day by day. It is estimated that 60%–80% of newly emerging or re-emerging viruses are zoonotic (animal-borne) viruses [7]. Any event that increases the likelihood of zoonotic viruses being transmitted to humans plays a role in the spread of these infections. The coexistence of different animal species in wet markets and the increase in the frequency of contact with them, the establishment of settlements in the habitats of animals, the increase in the frequency of working with different animal species in laboratories (by which we do not mean biological warfare studies but strictly normal scientific studies), climate change (mosquito movements) and the large increase in the international mobility of people (transport by plane, tourism, international sports events, commercial activities, and educational travel) are the most important reasons [7, 11]. However, viruses can also be useful tools. Viruses can be used as vectors for therapeutic purposes and in vaccine studies, to combat other viruses or bacterial infections resistant to antibiotics, such as bacteriophages [92]. Viruses can function like a small hand with which we can touch cells at the electron-microscope scale if used for good purposes; conversely, they have been used maliciously in the past by people who want to destroy humanity and are likely to be so used in the future. In the future, viral engineering studies may need to be subjected to worldwide inspection for reasons of of national and global security.

References and further reading

[1] Lodish H, Berk A, Zipursky S L, Matsudaira P, Baltimore D and Darnell J 2000 *Viruses: Structure, Function, and Uses (Molecular Cell Biology)* 4th edn (New York: W. H. Freeman) ch 6

[2] Pietilä M K, Demina T A, Atanasova N S, Oksanen H M and Bamford D H 2014 Archaeal viruses and bacteriophages: comparisons and contrasts *Trends Microbiol.* **22** 334–44

[3] Rodrigues R A L, Andrade A C, dos S P, Boratto P V, de M, Trindade G, de S, Kroon E G and Abrahão J S 2017 An anthropocentric view of the virosphere-host relationship *Front. Microbiol.* **30** 1673

[4] Woolhouse M, Scott F, Hudson Z, Howey R and Chase-Topping M 2012 Human viruses: discovery and emergence *Philos. Trans. R. Soc. B. Biol. Sci.* **367** 2864–71

[5] International Committee on Taxonomy of Viruses *Virus Taxonomy: 2019* (https://talk.ictvonline.org/taxonomy/) (Accessed, 4 March 2019)

[6] Krishnakumar V and Kannan M 2020 *Microbial Pathogenesis: Virus Pathogen–Host Interactions (Model Organisms for Microbial Pathogenesis, Biofilm Formation and Antimicrobial Drug Discovery)* ed B Siddhardha, M Dyavaiah and A S Syed (Singapore: Springer) ch 2
[7] Morens D M and Fauci A S 2013 Emerging infectious diseases: threats to human health and global stability *PLoS Pathog.* **9** 1003467
[8] Amor S 2009 *Virus Infections of the Central Nervous System (Manson's Tropical Diseases)* ed G C Cook and A I Zumla (London: Elsevier) ch 48
[9] Şahiner F 2016 Global spread of Zika virus epidemic: current knowledges and uncertainties *Mikrobiyol Bul.* **50** 333–51
[10] McEntire C R S *et al* 2021 Neurologic manifestations of the World Health Organization's list of pandemic and epidemic diseases *Front. Neurol.* **22** 634827
[11] van Aken J and Hammond E 2003 Genetic engineering and biological weapons *EMBO. Rep.* **4** 57–60
[12] Pohanka M 2019 Current trends in the biosensors for biological warfare agents assay *Materials. (Basel)* **12** 12142303
[13] Şahiner F, Siğ A K, Savaşçi Ü, Tekin K and Zika A F 2017 Virus-associated ocular and neurologic disorders *Pediatr. Infect. Dis. J.* **36** 341–46
[14] Sahiner F *et al* 2015 Congenital cytomegalovirus infections and glycoprotein B genotypes in live-born infants: a prevalence study in Turkey *Infect. Dis.* **47** 465–71
[15] Pretorius M and Venter M 2017 *Diagnosis of Viral Infections (Viral Infections in Children, Volume I)* (Cham: Springer) ed R J Greeen ch 6
[16] Christaki E 2015 New technologies in predicting, preventing and controlling emerging infectious diseases *Virulence* **6** 558–65
[17] Şahiner F and Gümral R 2020 Hepatit C infeksiyonlarının laboratuvar tanısı, karşılaşılan güçlükler ve güncel tanı algoritması *Flora J. Infect. Dis. Clin. Microbiol.* **25** 139–53
[18] Saylan Y, Erdem Ö, Ünal S and Denizli A 2019 An alternative medical diagnosis method: biosensors for virus detection *Biosensors* **9** 9020065
[19] Hoşbul T, Şahiner F and Sarı O 2020 Basic epidemiological parameters at the end of the 5th month of the COVID-19 outbreak *J. Mol. Virol. Immunol.* **1** 67–80
[20] Şahiner F and Hoşbul T 2020 Genomic and biologic characteristics of SARS-CoV-2 and other Coronaviruses *J. Mol. Virol. Immunol.* **1** 18–29
[21] Gorbalenya A E *et al* 2020 The species severe acute respiratory syndrome-related coronavirus: classifying 2019-nCoV and naming it SARS-CoV-2 *Nat. Microbiol.* **5** 536–44
[22] Wevers B A and van der Hoek L 2009 Recently discovered human Coronaviruses *Clin. Lab. Med.* **29** 715–24
[23] World Health Organization (WHO), Geneva, Switzerland *Summary of Probable SARS Cases with Onset of Illness from November 1, 2002 to July 31, 2003* (https://who.int/publications/m/item/summary-of-probable-sars-cases-with-onset-of-illness-f)
[24] World Health Organization (WHO) *Middle East Respiratory Syndrome Coronavirus (MERS-CoV)* (http://emro.who.int/pandemic-epidemic-diseases/mers-cov/mers-situation-update-june-2019.html)
[25] Shaw K 2006 The 2003 SARS outbreak and its impact on infection control practices *Public Health* **120** 8–14
[26] Zaki A M, van Boheemen S, Bestebroer T M, Osterhaus A D M E and Fouchier R A M 2012 Isolation of a novel Coronavirus from a man with pneumonia in Saudi Arabia *N. Engl. J. Med.* **367** 1814–20

[27] World Health Organization *WHO Coronavirus Disease (COVID-19) Dashboard* (https://covid19.who.int/)

[28] Şahiner F and Cebeci İ 2020 Major and possible transmission routes of SARS-CoV-2 infections *J. Mol. Virol. Immunol.* **1** 24–35

[29] Yılmaz Ü İ, Aydogan C N, Şahiner F and Hoşbul T 2021 mRNA Based SARS-CoV-2 vaccines and ongoing researches *J. Mol. Virol. Immunol.* **1** 19–29

[30] Erensoy S 2020 COVID-19 pandemisinde SARS-CoV-2 ve mikrobiyolojik tanı dinamikleri *Mikrobiyol. Bul.* **54** 497–509

[31] Şahiner F and Sümbül H E 2020 Rapid spreading of SARS-CoV-2 infection and risk factors: epidemiological, immunological and virological aspects *J. Mol. Virol. Immunol.* **1** 36–50

[32] Iravani S 2020 Nano- and biosensors for the detection of SARS-CoV-2: challenges and opportunities *Mater. Adv.* **1** 3092–103

[33] Centers for Disease Control (CDC) 1981 Pneumocystis pneumonia-Los Angeles *MMWR. Morb. Mortal. Wkly. Rep.* **30** 250–52

[34] Sepkowitz K A 2001 AIDS – the first 20 years *N. Engl. J. Med.* **344** 1764–72

[35] Centers for Disease Control (CDC) 1982 Update on acquired immune deficiency syndrome (AIDS) – United States *MMWR. Morb. Mortal. Wkly. Rep.* **31** 507–14

[36] Sharp P M and Hahn B H 2011 Origins of HIV and the AIDS pandemic *Cold. Spring. Harb. Perspect. Med. Sep* **1** 006841

[37] Kamps B S and Hoffmann C 2007 *Introduction (HIV Medicine)* ed C Hoffmann, J K Rockstroh and B Kamps 14th edn (Paris: Flying Publisher) ch 1

[38] Pillay D, Geretti A M and Weiss R A 2009 *Human-Immunodeficiency Viruses (Principles and Practice of Clinical Virology)* ed A J Zuckerman, J E Banatvala, B D Schoub, P D Griffiths and P Mortimer 6th edn (Oxford: Wiley) ch 38

[39] Berman D and Scott G B 2006 *Human Immunodeficiency Virus Type 1 (Congenital and Perinatal Infections)* ed C Hutto (Totowa: Humana Press) ch 14

[40] Boldogh I, Albrecht T and Porter D D 1996 *Persistent Viral Infections (Medical Microbiology)* ed S Baron 4th edn (Galveston, TX: University of Texas Medical Branch at Galveston) ch 46

[41] The Joint United Nations Programme on HIV and AIDS (UNAIDS), Geneva, Switzerland *UNAIDS Data 2020* (https://unaids.org/en/resources/documents/2020/unaids-data)

[42] Altfeld M and Walker B D 2007 *Acute HIV-1 Infection (HIV Medicine)* ed C Hoffmann, J K Rockstroh and B S Kamps 14th edn (Paris: Flying Publisher) ch 2

[43] Hong F, Aga E, Cillo A R, Yates A L, Besson G, Fyne E, Koontz D L, Jennings C, Zheng L and Mellorset J W 2016 Novel assays for measurement of total cell-associated HIV-1 DNA and RNA *J. Clin. Microbiol.* **54** 902–11

[44] Lu H and Tang Y-W 2019 Myths in the laboratory diagnosis of HIV infection *Emerg. Microbes. Infect.* **8** 1240–42

[45] Nguyen N L, Arcenas R and Tang Y-W 2018 *Laboratory Diagnosis of HIV-1 Infections: State of the Art (Advanced Techniques in Diagnostic Microbiology)* ed Y-W Tang and C W Stratton 4th edn (Switzerland: Springer) ch 18

[46] Wang L, Wang J, Tian X, Ruan J, Yu Y and Yan F 2019 Sample-to-cutoff ratios using architect HIV Ag/Ab combo: the influence with the results of supplemental tests and optimal cutoff value to predict HIV infection *J. Clin. Lab. Anal.* **33** 22866

[47] Decker A, Neumann-Haefelin C and Thimme R 2021 Virushepatitis A–E *Internist. (Berl).* **62** 163–74

[48] Cebeci İ and Şahiner F 2020 Hepatitis C virus: genetic characteristics, advances and current challenges for vaccine development *J. Mol. Virol. Immunol.* **1** 1–13
[49] World Health Organization (WHO) *Hepatitis* (https://who.int/health-topics/hepatitis#tab=tab_1)
[50] Fung J, Lai C L and Yuen M F 2009 Hepatitis B and C virus-related carcinogenesis *Clin. Microbiol. Infect.* **15** 964–70
[51] Jefferies M, Rauff B, Rashid H, Lam T and Rafiq S 2018 Update on global epidemiology of viral hepatitis and preventive strategies *World J. Clin. Cases* **6** 589–99
[52] Villar L M, Cruz H M, Barbosa J R, Bezerra C S, Portilho M M, Scalioni L and de P 2015 Update on hepatitis B and C virus diagnosis *World J. Virol.* **4** 323–42
[53] Sahiner F and Ardic N 2016 An update on Ebola virus epidemiology and experimental modalities *Dis. Mol. Med.* **4** 43–50
[54] Miranda M E *et al* 1999 Epidemiology of Ebola (subtype Reston) virus in the Philippines, 1996 *J. Infect. Dis.* **179** 115–19
[55] Tseng C-P and Chan Y-J 2015 Overview of Ebola virus disease in 2014 *J. Chinese. Med Assoc.* **78** 51–5
[56] Gatherer D The 2014 Ebola virus disease outbreak in West Africa *J. Gen. Virol.* **95** 1619–24
[57] Andreas A, Egom C B A, Kruzliak P and Egom E E 2015 Is there a way out for the 2014 Ebola outbreak in Western Africa? *Asian. Pac. J. Trop. Med.* **8** 773–78
[58] Feldmann H and Geisbert T W 2011 Ebola haemorrhagic fever *Lancet* **377** 849–62
[59] Matua G A, Van der Wal D M and Locsin R C 2015 Ebolavirus and haemorrhagic syndrome *Sultan. Qaboos. Univ. Med. J.* **15** 26052448
[60] Buathong R *et al* 2015 Detection of Zika virus infection in Thailand, 2012–2014 *Am. J. Trop. Med. Hyg.* **93** 380–83
[61] Staples J E *et al* 2016 Interim guidelines for the evaluation and testing of infants with possible congenital Zika virus infection – United States *MMWR. Morb. Mortal. Wkly. Rep.* **65** 63–7
[62] Ioos S, Mallet H-P, Leparc Goffart I, Gauthier V, Cardoso T and Herida M 2014 Current Zika virus epidemiology and recent epidemics *Médecine. Mal. Infect.* **44** 302–7
[63] Attar N 2016 ZIKA virus circulates in new regions *Nat. Rev. Microbiol.* **14** 2015.28
[64] Oehler E, Watrin L, Larre P, Leparc-Goffart I, Lastère S, Valour F, Baudouin L, Mallet H P, Musso D and Ghawche F 2014 Zika virus infection complicated by Guillain-Barré syndrome – case report, French Polynesia, December 2013 *Euro. Surveill.* **19** 20720
[65] Schuler-Faccini L *et al* 2016 Possible association between Zika virus infection and microcephaly – Brazil, 2015 *MMWR. Morb. Mortal. Wkly. Rep.* **65** 59–62
[66] Oliveira Melo A S, Malinger G, Ximenes R, Szejnfeld P O, Alves Sampaio S and Bispo de Filippis A M 2016 Zika virus intrauterine infection causes fetal brain abnormality and microcephaly: tip of the iceberg? *Ultrasound. Obstet. Gynecol.* **47** 6–7
[67] Dick G W, Kitchen S and Haddow A 1952 Zika virus (I) Isolations and serological specificity *Trans. R. Soc. Trop. Med. Hyg.* **46** 509–20
[68] Fauci A S and Morens D M 2016 Zika virus in the Americas – yet another arbovirus threat *N. Engl. J. Med.* **374** 601–4
[69] Besnard M, Lastere S, Teissier A and Cao-Lormeau V M D 2014 Evidence of perinatal transmission of Zika virus, French Polynesia, December 2013 and February 2014 *Euro. Surveill.* **19** 20751
[70] Foy B D, Kobylinski K C, Foy J L C, Blitvich B J, Travassos da Rosa A, Haddow A D, Lanciotti R S and Tesh R B 2011 Probable non–vector-borne transmission of Zika virus, Colorado, USA *Emerg. Infect. Dis.* **17** 880–2

[71] Shapshak Z, Somboonwit P, Foley C, Alrabaa B T, Wills S F and Sinnott J T T 2015 *Zika Virus (Global Virology I- Identifying and Investigating Viral Diseases)* ed P Shapshak, J T Sinnott, C Somboonwit and J Kuhn 1st edn (New York: Springer) ch 18

[72] Dupont-Rouzeyrol M, O'Connor O, Calvez E, Daurès M, John M and Grangeon J P 2014 Co-infection with Zika and Dengue viruses in two patients, New Caledonia *Emerg. Infect. Dis.* **21** 381–2

[73] Musso D, Nhan T, Robin E, Roche C, Bierlaire D, Zisou K, Shan Yan A, Cao-Lormeau V M and Broult J 2014 Zika virus transmission through blood transfusion demonstrated during an outbreak in French Polynesia, November 2013 to February 2014 *Euro. Surveill.* **19** 20761

[74] Robilotti E, Deresinski S and Pinsky B A 2015 Norovirus *Clin. Microbiol. Rev.* **28** 134–64

[75] International Committee on Taxonomy of Viruses (ICTV) *Caliciviridae* (https://talk.ictvonline.org/ictv-reports/ictv_9th_report/positive-sense-rna-viruses-2011/w/posrna_viruses/253/caliciviridae)

[76] Rothemund P W K 2004 Norovirus cultured for the first time 2004 *PLoS Biol.* **2** e445

[77] Kapikian A Z 2000 The discovery of the 27-nm Norwalk virus: an historic perspective *J. Infect. Dis.* **181** 295–302

[78] de Graaf M, van Beek J and Koopmans M P G 2016 Human norovirus transmission and evolution in a changing world *Nat. Rev. Microbiol.* **14** 421–33

[79] Centers for Disease Control and Prevention (CDC) *Norovirus* (https://cdc.gov/norovirus/index.html)

[80] Marks P J, Vıpond I B, Carlısle D, Deakın D, Fey R and Caul E O 2000 Evidence for airborne transmission of Norwalk-like virus (NLV) in a hotel restaurant *Epidemiol. Infect.* **124** 481–7

[81] Toennessen R, Lauscher A and Rimstad E 2009 Comparative aspects of infectious salmon anemia virus, an orthomyxovirus of fish, to influenza viruses *Indian. J. Microbiol.* **49** 308–14

[82] International Committee on Taxonomy of Viruses (ICTV) *Orthomyxoviridae* (https://talk.ictvonline.org/ictv-reports/ictv_9th_report/negative-sense-rna-viruses-2011/w/negrna_viruses/209/orthomyxoviridae)

[83] Krammer F *et al* 2018 Influenza *Nat. Rev. Dis. Prim.* **4** 3

[84] Mattio L M, Catinella G, Pinto A and Dallavalle S 2020 Natural and nature-inspired stilbenoids as antiviral agents *Eur. J. Med. Chem.* **202** 112541

[85] World Health Organization (WHO) *Influenza* (http://who.int/news-room/fact-sheets/detail/influenza-(seasonal))

[86] Bui C H T *et al* 2019 Tropism of influenza B viruses in human respiratory tract explants and airway organoids *Eur. Respir. J.* **54** 1900008

[87] Meşe S, Uyanik A, Özakay A, Öztürk S and Badur S 2018 Influenza surveillance in Western Turkey in the era of quadrivalent vaccines: a 2003–2016 retrospective analysis *Hum. Vaccin. Immunother.* **14** 1899–908

[88] Hay A J, Gregory V, Douglas A R and Lin Y P 2001 The evolution of human influenza viruses *Philos. Trans. R. Soc. London. Ser. B. Biol. Sci.* **356** 1861–70

[89] Centers for Disease Control and Prevention (CDC) *Influenza C* (https://cdc.gov/flu/index.html)

[90] Henderson D A 2011 The eradication of smallpox – An overview of the past, present, and future *Vaccine.* **29** D7–9

[91] Saha A, Kaul R, Murakami M and Robertson E S 2010 Tumor viruses and cancer biology: modulating signaling pathways for therapeutic intervention *Cancer. Biol. Ther.* **10** 961–78

[92] Barbu E M, Cady K C and Hubby B 2016 Phage therapy in the era of synthetic biology *Cold Spring Harb. Perspect. Biol.* **8** 023879

Chapter 2

Optical biosensors for virus detection

Yeşeren Saylan and Adil Denizli

The coverage of biosensors in research literature has experienced excellent growth in the last 20 years, with a huge number of studies on different platforms. Herein, we describe novel and updated studies of the use of optical biosensors for the analysis of viral infections. This chapter aims to provide fundamental knowledge in the field of optical biosensors, including their preparation methods, properties, and characterization techniques, along with the state-of-the-art technological developments that incorporate other platforms such as microfluidic systems, lab-on-a-chip tools, and molecular imprinting techniques. It covers the importance of optical biosensors, in particular, for different viruses such as coronavirus, human immunodeficiency virus, hepatitis, Ebola, Zika, norovirus, influenza, and others. This chapter also covers the integration of optical biosensors with current top-notch tools and platforms to ultimately facilitate their potential application in the realm of medicine.

2.1 Introduction

Viruses are small entities with diameters between 20 and 400 nm that can be rapidly transmitted between people by contact with droplets, aerosols, and even fecal matter [1, 2]. Viral diseases have been a huge threat to community stability. An urgent requirement for efficient control of present and future threats has forced the development of sensitive, accurate, and fast monitoring and detection platforms [3–8]. Virus diagnoses have depended on culture-based and serological methods for direct (intact viruses and their components) and indirect (antibodies and antigens) methods [9]. The development of these methods for the diagnosis of infectious diseases has allowed the quantitative reverse-transcriptase polymerase chain reaction and the polymerase chain reaction to become the standard methods for the diagnosis of several common viruses [10–13]. Despite the benefits of these traditional methods in terms of selectivity and sensitivity, the requirement for time-consuming steps, the need for huge equipment, and the large number of false negative results limit the efficiency of these approaches for avoiding new infection risks [14]. Thus, the need for rapid, real-time, and low-cost

doi:10.1088/978-0-7503-3867-7ch2 © IOP Publishing Ltd 2021

methods has moved the focus to the development of biosensor platforms. Biosensors include various transducers and recognition elements [15], which capture target molecules and convert their reactions into measurable signals. They can be categorized as optical [16], electrochemical [17], piezoelectric [18], thermal [19], or magnetic [20]. Among this diversity of biosensors, optical biosensors have attracted substantial interest due to their adaptability, label-free sensing, and short response time. They have numerous subclasses, depending on whether they measure reflection, adsorption, refractive index, Raman spectra, fluorescence, infrared spectra, chemiluminescence, or energy transfer [21–26].

Herein, the main principle of optical biosensors is clarified, the importance of optical biosensors in medical diagnosis and the use of optical biosensors are extensively examined, and finally, conclusions and future perspectives are offered at the end of the chapter.

2.2 Principle of optical biosensors

Analytical systems comprise a recognition element and a transducer, which detect a target molecule for sample analysis and then estimate the combination, construction, and function of target molecules by adapting their responses to the signals [27]. The target molecule associates with the receptor, and the detection part determines the presence of a target molecule using a binding reaction. The transducer transforms the alterations to the signal measured by the detector [28]. As depicted in figure 2.1, an ideal biosensor offers multiple properties including selectivity and

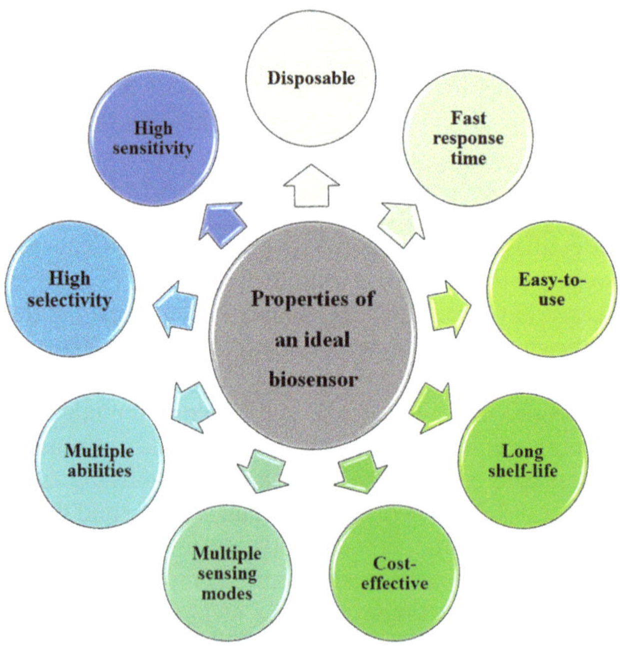

Figure 2.1. Properties of an ideal biosensor.

sensitivity, quick response time, disposability, ease of use, long shelf life, cost-effectiveness, and a multiple-sensing capability [29]. Optical biosensor development has attracted much attention in recent years [30]. Optical biosensors evaluate changes in the reflection, absorption, refractive index, Raman spectra, fluorescence, infrared spectrum, chemiluminescence, or energy transfer when the recognition element and target molecule form the complex [31–40]. Researchers have continued to improve biosensor sensitivity by focusing on fabrication quality, amplifying the affinity between different surface chemistry modifications, and using polymers for signal amplification studies [41–45].

2.3 Importance of optical biosensors in medical diagnosis

Optical biosensor research began several years ago, and numerous scientists, including biologists, biochemists, chemists, biophysicists, physicists, bioengineers, and medical doctors have been using these biosensors in innovative applications in several disciplines, such as medical diagnosis [46], environmental studies [47], pharmaceutics [48], and so on. Among these, medical diagnosis has been investigated as an effective application field because diseases, especially infectious diseases, still present a universal danger to human health [49, 50]. Important maladies can spread because of the lack of accessible methods, varying treatment approaches, and fragmentary access to well-equipped health facilities [51]. Due to the requirements for enhanced monitoring features and rapid analysis, different recognition elements and platforms have been applied to recognition. The detection performance of optical biosensors can be improved by using them in different combinations. These features make them appropriate for medical diagnosis, supporting multi-molecule and rapid monitoring [52–55].

2.4 Latest findings in the field of optical biosensors

Viruses are infective entities and they cause serious disease in populations. According to statistics, viruses are the global cause of more than two million deaths per year, with the exception of coronaviruses [56–59]. In this section, the latest findings in the field of optical biosensors for the detection of different viruses are exemplified in detail.

2.4.1 Coronavirus

Qiu *et al* built a dual-mode optical biosensor for coronavirus detection. This biosensor used a combination of localized surface plasmon resonance and plasmonic photothermal effects and was made from gold nanomaterials on a glass surface. The nanomaterial was modified using complementary DNA receptors to conduct the detection. Qiu *et al* allowed the sample to fall at two different angles to improve the sensing features and calculated the detection limit to be 0.22 pM [60]. Moitra *et al* described a colorimetric optical biosensor using nanoparticles functionalized with thiol-based oligonucleotides for coronavirus detection. Their sensing system was specific to the nucleocapsid protein from RNA samples and was also employed to detect coronavirus in a short time (10 min). The nanoparticle-based biosensor collected the RNA sequence showing the presence of coronavirus, which led to a shift in the spectrum due to the plasmon effect. They also investigated the selectivity of the biosensor against Middle East respiratory syndrome (MERS) RNA and determined the limit of detection value to be 0.18 ng μL^{-1} in the 0.2–3 ng μL^{-1} concentration range [61]. Seo *et al* proposed an optical biosensor based on a field-effect transistor for coronavirus detection. They coated the biosensor surface with graphene sheets and an antibody for coronavirus spike protein. The coronavirus spike antibody was coated on the graphene sheet using a linker. They performed experiments using antigen protein, cultured virus, and swab specimens from virus-positive patients. The optical biosensor was able to detect the coronavirus antigen protein at concentrations as low as (1 fg mL^{-1}). They mentioned that the biosensor detected the antigens in medical samples without any processing and that the biosensor also had no cross-reactivity with the MERS virus antigen [62]. Liu *et al* reported a surface-enhanced Raman scattering (SERS)-dependent lateral flow assay-based optical biosensor designed to sense the immunoglobulins (IgM and IgG) of coronavirus. They labeled two layers of dye via a silver shell on a silicon dioxide core coating and the biosensor showed outstanding signals that were well monodispersed, with high uniformity. They modified the strips with anti-human IgM and IgG to detect spike protein anti-IgM/IgG complexes (figure 2.2) and obtained a detection limit 800 times higher than that of a standard gold nanoparticle-based strip. They also used their sensor with nineteen samples from positive patients and forty-nine negative samples from healthy persons to show the medical practicability of biosensors [63]. Huang *et al* developed an optical biosensor that required no extra preparation steps for spike-protein-specific detection (figure 2.3). The biosensors detected a low concentration of 370 vp mL^{-1} in one step in a short time in the range

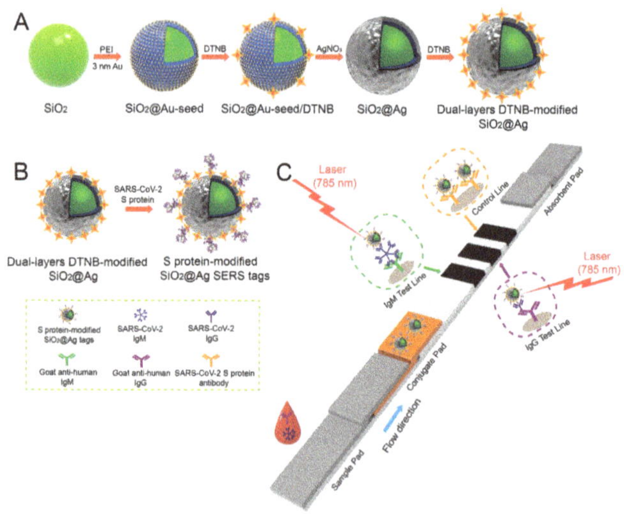

Figure 2.2. (A) Schematic of the preparation of the layers of dye on a nanoparticle, (B) coronavirus-protein-functionalized SERS tag preparation, and (C) the analysis of IgM/IgG by the strip. Reprinted from [63], Copyright 2021, with permission from Elsevier.

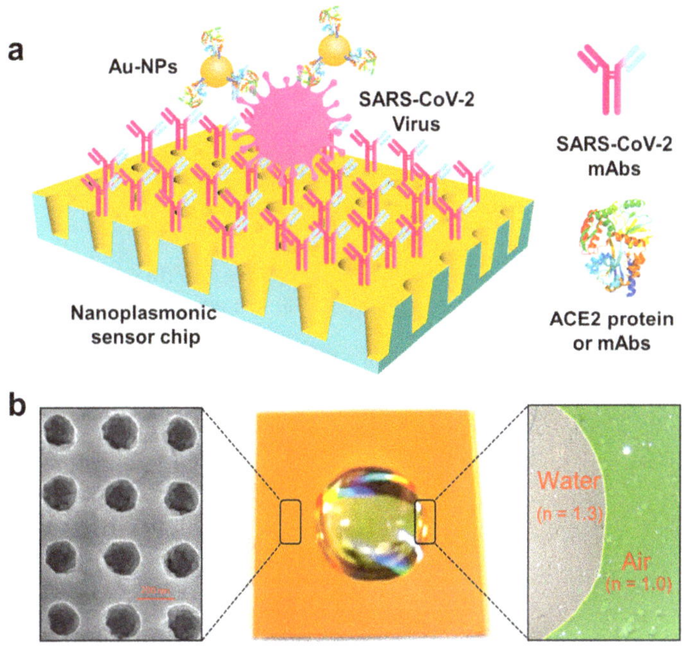

Figure 2.3. (a) Schematic of the optical biosensor developed for coronavirus detection and (b) optical photo (middle) of a gold chip piece with a water drop, scanning electron microscopy image (left), transmission microscopy image (right). Reprinted from [64], Copyright 2021, with permission from Elsevier.

of 0 to 10^7 vp mL^{-1}. They connected a smartphone to demonstrate a rapid and low-cost detection method that can be applied in both medical and resource-limited settings [64].

2.4.2 Human immunodeficiency virus

Diao *et al* prepared an optical biosensor that depended on a surface plasmon resonance sensing strategy to detect HIV-related DNA using entropy-driven strand displacement interactions and two-layer DNA tetrahedrons. The entropy-driven strand displacement reactions used for enzyme- and label-free amplification can be initiated by DNA, guided to DNA cyclic utilization, and form double-stranded DNA. These DNA parts then interact with the immobilized capture probes and associate with the DNA tetrahedral nanostructures. Owing to the high efficacy of the reactions and the large molecular weight of the double-layer DNA tetrahedrons, the signal was dramatically enhanced, as depicted in figure 2.4. The optical biosensor offered detection in a broad range from 1 pM to 150 nM with a low limit of detection value of (48 fM) [65]. Fu *et al* reported a study of a SERS-based and lateral flow assay-based biosensor used to detect HIV-1 DNA in low concentrations. Labeled gold nanoparticles were utilized as nanotags for the detection of HIV-1 DNA (figure 2.5). They mentioned that the biosensor could quantitatively examine HIV-1 DNA by sensing the peak intensity of the gold nanoparticles. Under optimized conditions, they calculated that the detection limit was 0.24 pg mL^{-1},

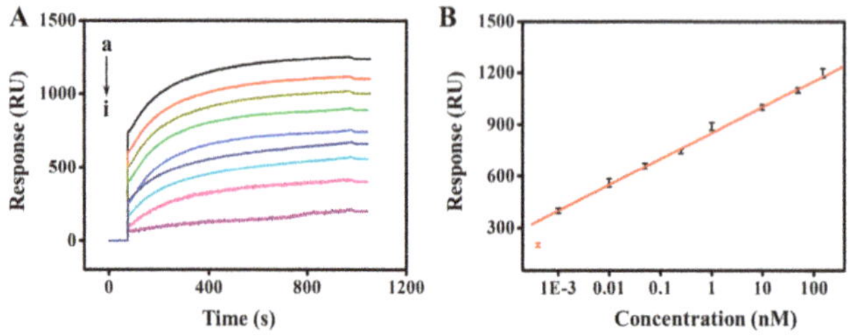

Figure 2.4. (A) Sensorgrams for the detection of HIV-related DNA and (B) the calibration curve for HIV-related DNA detection. Reprinted from [65], Copyright 2018, with permission from Elsevier.

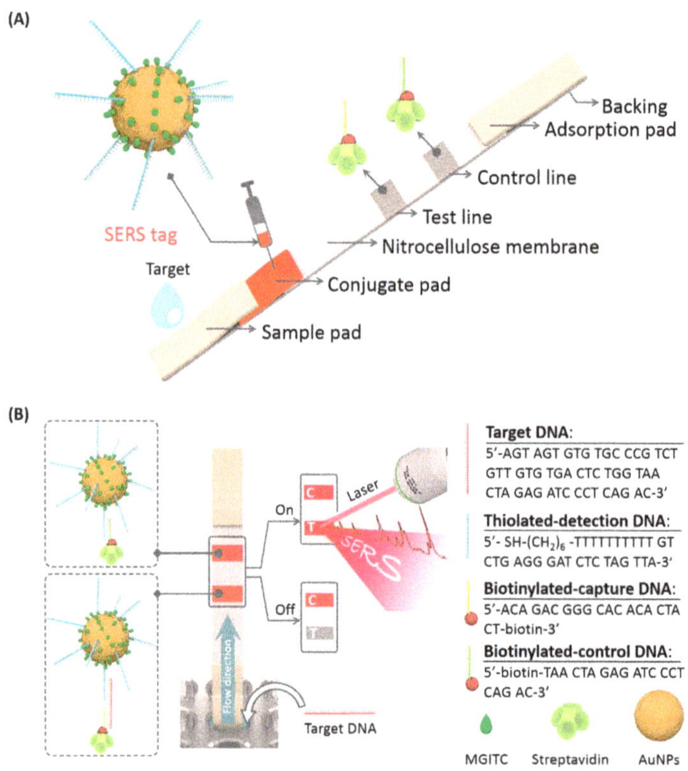

Figure 2.5. (A) Schematic of the detector and (B) principle of the biosensor designed for HIV-1 DNA detection. Reprinted from [66], Copyright 2016, with permission from Elsevier.

which was more sensitive than other fluorescent or colorimetric biosensors [66]. Kurdekar *et al* synthesized streptavidin-modified fluorescent silver nanoparticles for HIV-1 p24 antigen detection in medical samples. Concentrations of between 10 and 1000 pg mL^{-1} were detected using a dose-dependent method. They reported that false positive results were not noticed with plasma from healthy persons and that the results demonstrated no interference with antigen detection [67].

2.4.3 Hepatitis

Luo *et al* developed a multifunctional molecularly imprinted biosensor to detect the hepatitis A (HAV) and hepatitis B (HBV) viruses. They observed that optical detection was possible because of the red and green colors of quantum dots in different concentrations in the targets. They combined the interaction of hydrophilic monomers and metal chelation to decrease nonspecific interactions and improve the adsorption specificity (figure 2.6). They attained acceptable selectivity and sensitivity in the detection of both viruses with high imprinting factor values (3.70 and 3.35), and low limit of detection values (3.4 and 5.3 pmol L^{-1}) for HAV and HBV, respectively [68].

Sciuto *et al* prepared an optical biosensor using a silicon–plastic mixed technology that was able to control the extraction, purification, and detection of nucleic acid. They performed sample preparation via an extraction process that employed magnetic particles on a disk. Detection was completed by a quantitative real-time polymerase chain reaction. They carried out nucleic acid detection by applying the hepatitis B virus DNA genome. The quantitative real-time polymerase chain reaction system in the disk demonstrated an improvement in sensitivity. They calculated the detection and quantification limits to be around 8 and 26 cps/reaction,

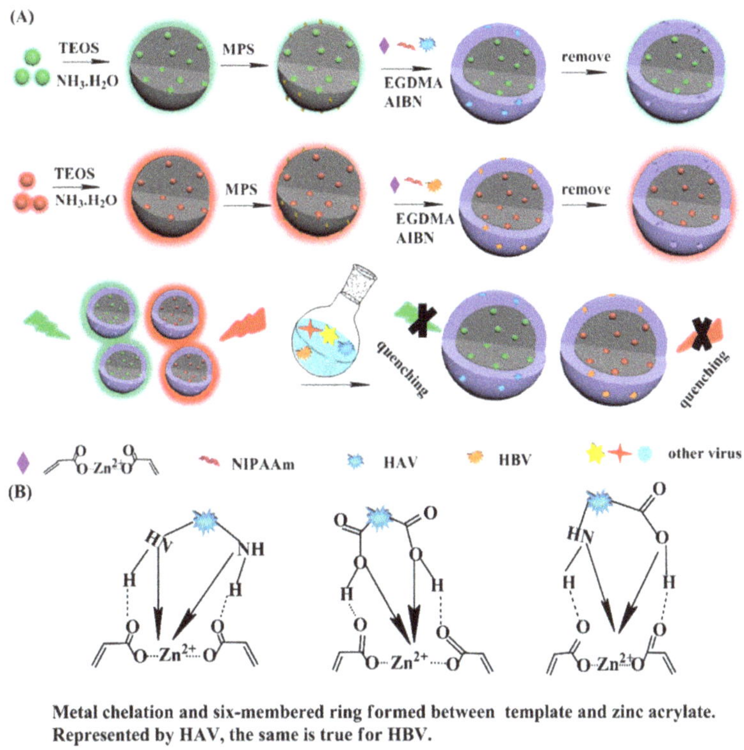

Figure 2.6. Scheme of the biosensor and (B) formation of metal chelation and ring. Reprinted with permission from [68]. Copyright 2019 American Chemical Society.

respectively, in analytical samples, while real sample results showed that the biosensor had high performance compared to the standard methods [69]. Kashif *et al* developed SERS to monitor hepatitis C virus (HCV) in blood serum samples. They used silver nanoparticles and obtained SERS spectra from positive and healthy individuals. They also evaluated the same sample sets with Raman spectroscopy and SERS and obtained more support for the detection of virus infection. As shown in figure 2.7, several SERS spectra associated with RNA bases are easily noticed in the positive sample when compared to those of healthy people, which can be used as spectral markers of the infection [70].

2.4.4 Ebola

Seymour *et al* prepared an interferometric reflectance imaging biosensor for multiple virus detection using the immobilization of a DNA-directed antibody. They developed three approaches for Ebola virus capture and detection (figure 2.8). The biosensor surface is marked with DNA sequences via the running conjugates of antibody DNA and permits the hybridization of DNA with DNA. The biosensor is able to sense three recombinant vesicular stomatitis viruses in real time, including Ebola. They also showed that the biosensor can be altered to generate a simple and homogeneous detection platform before incubation that eliminates the step of antibody immobilization. The biosensor was able to detect Ebola virus at a concentration of 100 PFU mL^{-1} [71].

Sharma *et al* screened different Ebola virus antibodies using a surface plasmon resonance biosensor. They first modified a gold biosensor surface with 4-mercapto-benzoic acid, and the modified surface was employed for the immobilization of the

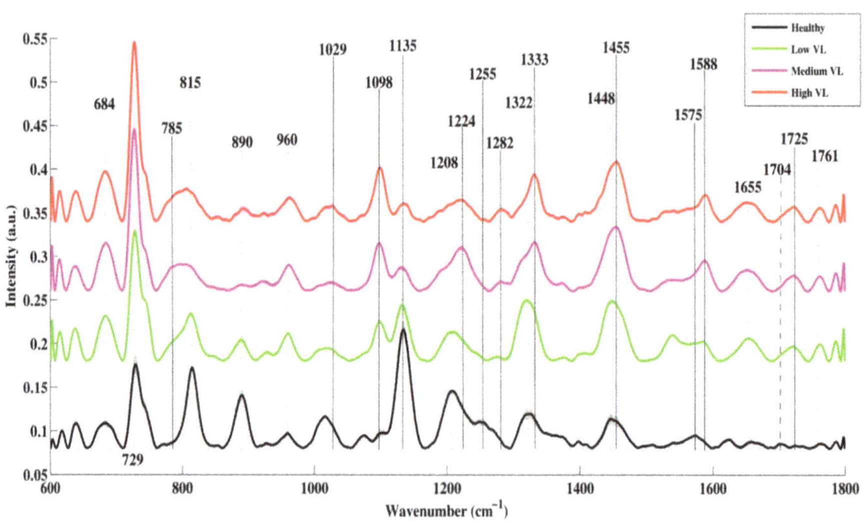

Figure 2.7. Spectra of SERS data from healthy (black) and positive samples with low (green), medium (magenta), and high (red) viral loads. Reprinted from [70], Copyright 2020, with permission from Elsevier.

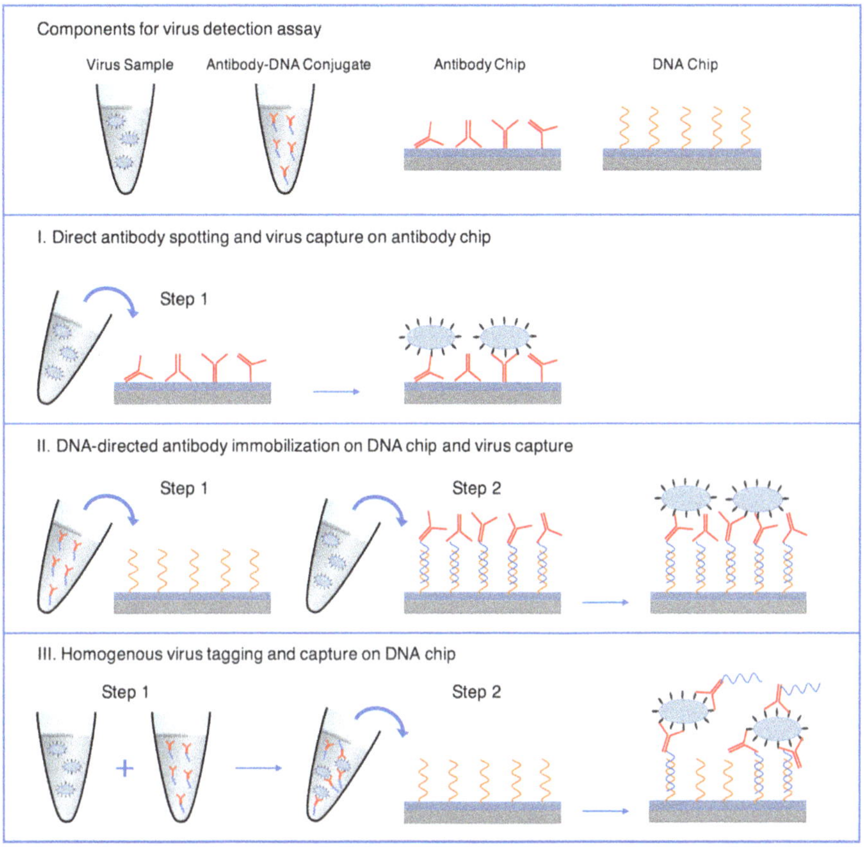

Figure 2.8. Different steps of Ebola virus detection. Reprinted with permission from [71]. Copyright 2021 American Chemical Society.

recombinant nucleoprotein of the Ebola virus. They investigated the antibody interactions to identify the best antibody according to the affinity constants and observed that mAb3 has the highest affinity with the lowest affinity constant (52 pM). They also measured the thermodynamic parameters of the interaction between the antibody and the recombinant nucleoprotein of the Ebola virus and reported that the binding was endothermic, spontaneous, and operated by entropy [72]. Natesan *et al* arranged a digital and multiplexed biosensor for the telemonitoring and detection of Ebola virus. The biosensor comprised a flow cell that conquered exact antibodies with antigens from all six species of virus and which used a smartphone fluorescent reader. They incorporated multiplexed viral proteins to get the most trustworthy results by cross-correlation and developed a custom smartphone application for data analysis, interpretation, and communication. The application checked the hardware attachment to turn on LEDs and recorded the optically enhanced images. They mentioned that the processing time was around 20 min for small amounts of blood, and that the results were processed and displayed within 15 seconds [73].

2.4.5 Zika

Draz *et al* reported a cell phone biosensor using nanomotor-based bead motion for Zika virus detection. The presence of the virus caused an accumulation of nanomotors on the bead surfaces, which generated motion (figure 2.9). They detected the Zika virus concentration using the motion alteration. The biosensor was able to detect Zika virus at very low limits (1 particle μL^{-1}). The biosensor also featured extremely precise detection of the presence of similar viruses. They pointed out that this biosensor had a high likelihood of being employed in point-of-care diagnostics developments [74].

Rong *et al* proposed a smartphone-based lateral flow biosensor for Zika virus nonstructural protein 1 detection. As demonstrated in figure 2.10, they designed an attachment and integrated a 3D-printed smartphone with external parts. They utilized quantum dots as probes in the biosensor, due to their very bright fluorescence signal. They found that the limits of detection values were 0.045 and 0.15 ng mL^{-1} in buffer and serum, respectively. They also performed cross-reactivity experiments with a similar protein, dengue virus nonstructural protein 1, and obtained a limited

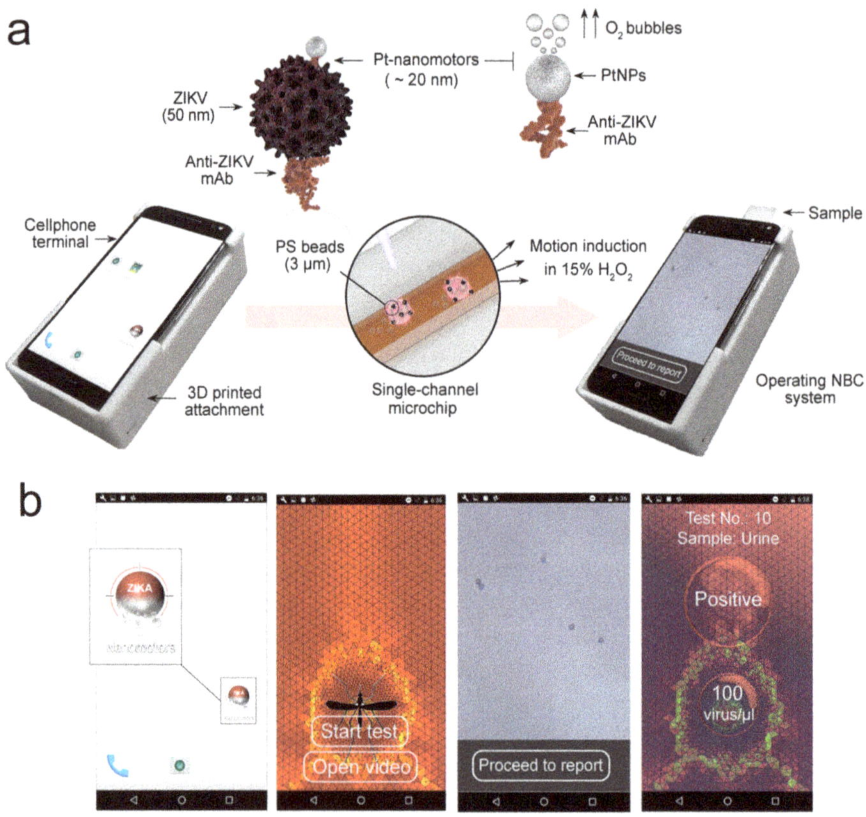

Figure 2.9. (a) Three parts of the biosensor's operation: (i) 3D-printed attachment, (ii) single-channel chip, and (iii) cell phone station; and (b) motion following and data handling. Reprinted with permission from [74]. Copyright 2018 American Chemical Society.

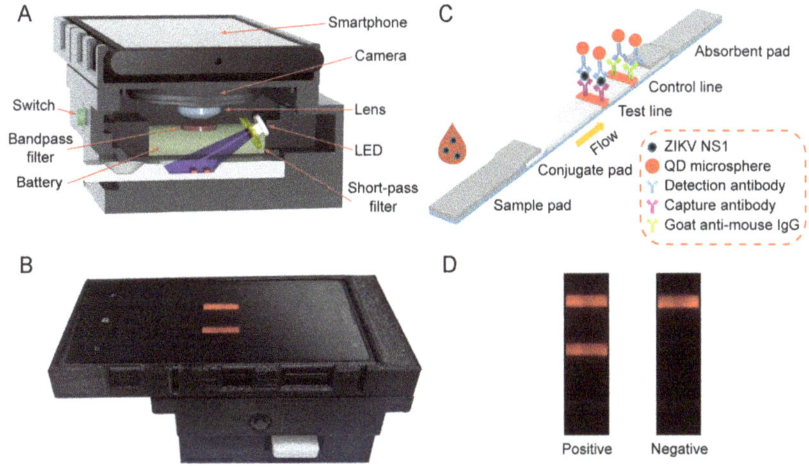

Figure 2.10. (A) 3D schematic of the smartphone-based biosensor, (B) optical photo of the fluorescent reader, (C) scheme for Zika virus nonstructural protein 1 detection, and (D) pictures of the strips in both the presence and absence of a target. Reprinted from [75], Copyright 2019, with permission from Elsevier.

interaction. Due to the biosensor's features, including compact size, low cost, and outstanding analytical performance, the fluorescence biosensor has significant potential for accurate and rapid Zika virus nonstructural protein 1 detection [75]. Zhang *et al* reported a biosensor based on a new class of Janus emulsion agglutination to detect protein–protein interactions, and hence, the Zika virus. They used Janus emulsion drops for the functionalization of an antigen-binding protein of the Zika virus to obtain a thermally stable system. The emulsion drops had dyes in the fluorocarbon and hydrocarbon phases, strengthening the signal, which was sensed via a simple optical fiber [76].

2.4.6 Norovirus

Weerathunge *et al* proposed a colorimetric biosensor to monitor norovirus. Their system used the enzyme-mimetic activity of gold nanoparticles and aptamers to prepare biosensors. The biosensor generated a blue color in the presence of norovirus and the color intensity changed with different concentrations of the virus (figure 2.11). They calculated the detection limit value to be 20 viruses/assay (200 viruses mL^{-1}) and also demonstrated the biosensor's stability via selectivity experiments in the existence of other similar microorganisms to demonstrate the ability of their sensor to detect norovirus in complex media [77].

Nasrin *et al* presented a localized surface plasmon resonance biosensor that combined fluorescent quantum dots with gold nanoparticles. The quantum dots covered with L-cysteine and citrate-stabilized gold nanoparticles were modified with 11-mercaptoundecanoic acid to produce functional groups on the surface (figure 2.12). They observed that the nanocomposite fluorescence had a quenched spectrum at 640 nm. Following the addition of norovirus-like particles, a steric-limitation-provoked signal from the nanoparticles triggered increased fluorescence in

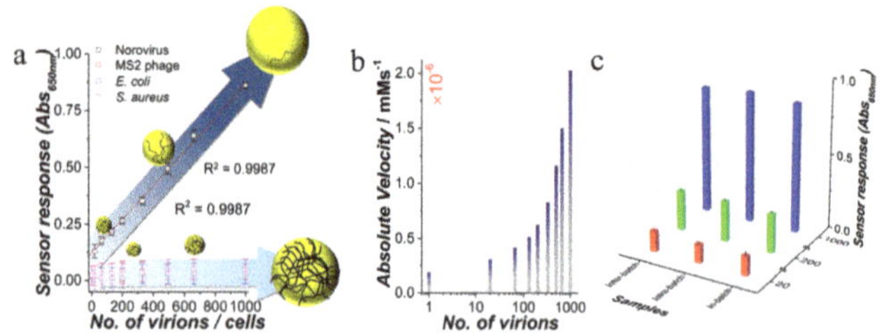

Figure 2.11. (a) Behavior of biosensor signals for an increasing number of entities, (b) reaction velocity in the absence and presence of different amounts of norovirus, and (c) in-, intra-, and inter-batch reproducibility of the biosensor for norovirus. Reprinted with permission from [77]. Copyright 2019 American Chemical Society.

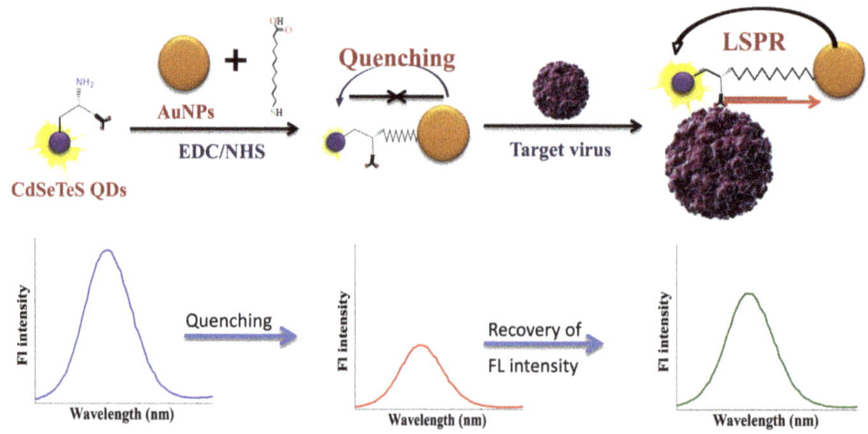

Figure 2.12. Scheme for the preparation of the nanocomposite and the norovirus detection mechanism. Reprinted from [78], Copyright 2018, with permission from Elsevier.

the dots. The detection limit value was found to be 12.1×10^{-15} g mL^{-1} in the 10^{-14} to 10^{-9} g mL^{-1} range [78]. Takemura *et al* developed a localized surface plasmon resonance-amplified biosensor to monitor norovirus in complex media. As shown in figure 2.13, they synthesized a multifunctional gold nanoparticle and magnetic nanoparticle hybrid nanocomposite and then conjugated it with the quantum dots using an antinorovirus antibody. This biosensor was able to separate the target virus from impurities using a magnetic field and induced a strong effect. This biosensor was able to detect norovirus-like particles in the broad range of 1 pg mL^{-1} to 5 ng mL^{-1} with a low detection limit value of 0.48 pg mL^{-1} in feces [79].

2.4.7 Influenza

Chen *et al* fabricated a nanomaterial-based optical biosensor that used a surface energy difference for influenza A/H1N1 virus detection. This difference controlled a gold nanoparticle self-assembly sequence and the nanoparticles produced various spots for detection. This biosensor was able to achieve an evaluation of the influenza

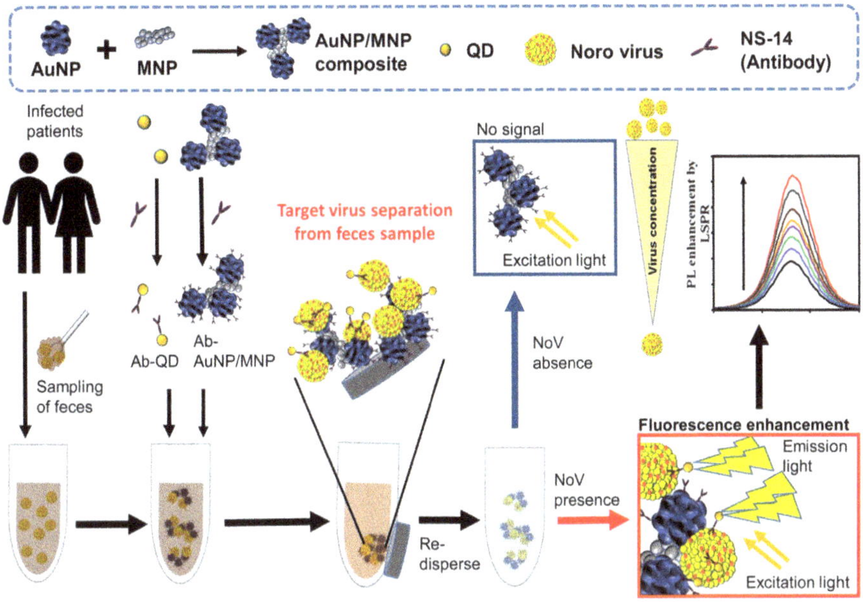

Figure 2.13. Schematic of the localized surface plasmon resonance biosensor designed for norovirus sensing. Reprinted from [79], Copyright 2019, with permission from Elsevier.

A/H1N1 virus using a Raman peak intensity decrease that resulted in a labeled aptamer DNA release (figure 2.14). The SERS-based biosensor had a detection limit of 97 PFU mL^{-1} and the time required was estimated to be 20 min [80]. Zhao *et al* reported a fiber-based surface plasmon resonance biosensor for avian influenza virus detection. They exposed the core surface and replaced it with a gold film to observe effective changes. They used a plasma alteration method to synthesize a monolayer on the optical fiber surface and calculated the detection limit to be 5.14×10^5 EID_{50}/ 0.1 ml after optimization experiments. Furthermore, they mentioned the advantages of the biosensor, such as low cost and compact size, which rendered it appropriate for online and remote sensing [81]. Lee *et al* prepared plasma magnetic graphene (PMGRP) using gold and iron oxide nanoparticles for a magnetofluoro immunosensing system for influenza virus A (H1N1) detection (figure 2.15). They obtained a sandwich arrangement between quantum dots and PMGRP by the conjugation of virus antibodies. The biosensor successfully detected a virus with a low limit value of (7.27 fg mL^{-1}) [82].

2.4.8 Other viruses

Peng *et al* demonstrated an ultrathin two-dimensional MXene nanomaterial-based optical biosensor designed to detect human papillomavirus. The biosensor exhibited a high quenching capability for labeled single-stranded DNA and diverse affinities for single-stranded and double-stranded DNA and had minimal fluorescent emission. Following the formation of a dual structure with the complementary target, the intensity clearly increased. The optical biosensor demonstrated a low limit of

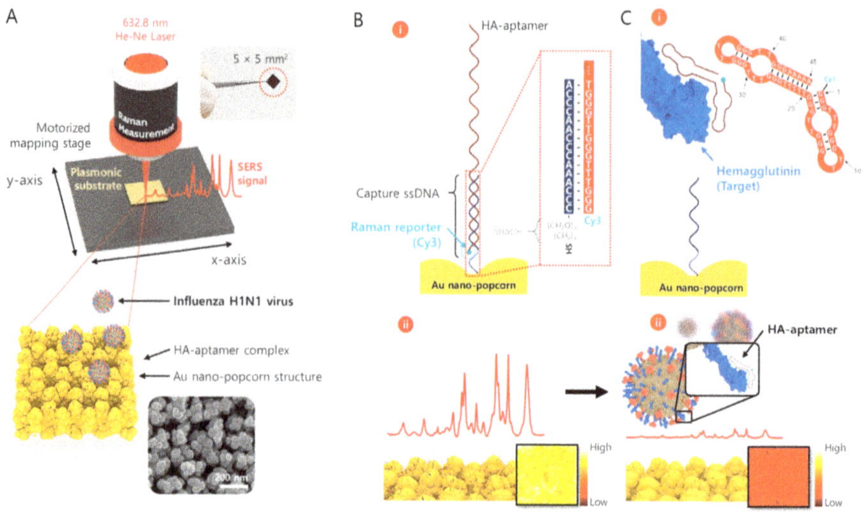

Figure 2.14. (A) Scheme of the SERS imaging-based biosensor for influenza A/H1N1 virus detection; (B) hybridization of labeled aptamer probes with capture DNA on the surface of the nanomaterial (i), creation of a signal (ii); (C) detection of virus encourages an aptamer change (i), guided signal intensity decline (ii). Reprinted from [80], Copyright 2020, with permission from Elsevier.

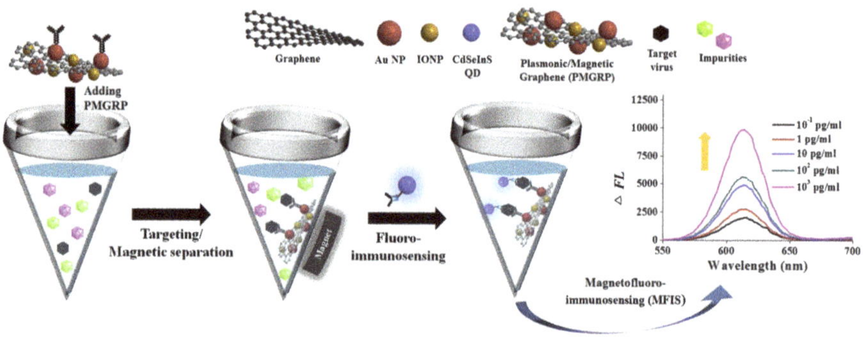

Figure 2.15. Schematic of virus detection using the magnetofluoro immunosensing process. Reprinted from [82], Copyright 2018, with permission from Elsevier.

detection value (100 pM) and was used to detect PCR-amplified human papillomavirus in cervical samples [83]. Guerreiro *et al* improved a genetically-encoded fluorescent biosensor using a green fluorescent protein with a cleavable spot for adenovirus detection. Following the optimization experiments, the biosensor was validated by expression in mammalian cells and employed for live-cell adenovirus sensing. They reported that the biosensor was triggered at 48 h post infection, had a detection limit of 10^5 particles mL^{-1}, and was notably more rapid than standard plaque-forming assays [84]. Teengam *et al* studied an optical biosensor used to screen for MERS, human papillomavirus, and *Mycobacterium tuberculosis*. They developed a DNA-related multiplex paper-dependent system that depended on peptide nucleic acid probes to obtain stable, easy-to-synthesize, and efficient hybridization. Detection was achieved by quantifying nanoparticles' color change, and the detector achieved limit of detection values of 1.53, 1.03, and 1.27 nM for MERS coronavirus, human papillomavirus, and *Mycobacterium tuberculosis*, respectively [85]. In our final example, Kim *et al* proposed a colorimetric biosensor that depended on double-stranded DNA-covered gold nanoparticles for MERS detection. This biosensor was capable of verifying the presence of virus owing to the shift and color changes of gold nanoparticles and was able to distinguish concentrations as low as 1 pmol μL^{-1} [86].

2.5 Conclusions and future perspectives

Medical application systems have constituted an enormous research field in recent years, but they still have to address numerous unmet conditions that are essential for the advancement of new sensing systems. Biosensors should be rapid, stable, reliable, simple, disposable, and capable of multiple sensing in medical applications. The use of optical biosensors has started to leave the proof-of-concept stage, and several viruses have been monitored to demonstrate the properties of biosensors. Sometimes, real sample validation experiments in medical applications strengthen optical biosensors' applicability. There are some hesitations to overcome. The rate of development is not fast, but a noteworthy progression is the use of smartphones for mobile health diagnostic applications. An exponential increase in the growth of these applications is required to revolutionize health delivery. They will open a novel window onto worldwide health access.

References

[1] Mauriz E 2020 Recent progress in plasmonic biosensing schemes for virus detection *Sensors* **20** 4745

[2] Vermisoglou E, Panacek D, Jayaramulu K, Pykal M, Frebort I, Kolar M, Hajdúch M, Zboril R and Otyepka M 2020 Human virus detection with graphene-based materials *Biosens. Bioelectron.* **166** 112436

[3] Ehtesabi H 2020 Application of carbon nanomaterials in human virus detection *J. Sci-Adv. Mater. Dev.* **5** 436–50

[4] Erdem Ö, Derin E, Sagdic K, Yilmaz E G and Inci F 2021 Smart materials-integrated sensor technologies for COVID-19 diagnosis *Emerg. Mater.* **4** 169–85

[5] Artika I M, Wiyatno A and Maroef C N 2020 Pathogenic viruses: molecular detection and characterization *Infect. Genet. Evol.* **81** 104215

[6] Abid S A, Suhail A, Al-Kadmy I M S, Sattar A A, Beshbishy A M, El-Saber Batiha G and Hetta H F 2018 Biosensors as a future diagnostic approach for COVID-19 *Life Sci.* **143** 359–73

[7] Qin Z, Peng R, Baravik I K and Liu X 2020 Fighting COVID-19: integrated micro- and nanosystems for viral infection diagnostics *Matter* **3** 628-51

[8] Saylan Y, Erdem Ö, Ünal S and Denizli A 2019 An alternative medical diagnosis method: biosensors for virus detection *Biosensors* **9** 65

[9] Katsarou K, Bardani E, Kallemi P and Kalantidis K 2019 Viral detection: past, present, and future *BioEssays* **41** 1900049

[10] Bhalla N, Pan Y, Yang Z and Payam A F 2020 Opportunities and challenges for biosensors and nanoscale analytical tools for pandemics: COVID-19 *ACS Nano* **14** 7783–807
[11] Pishva P and Yüce M 2021 Nanomaterials to tackle the COVID-19 pandemic *Emerg. Mater.* **4** 211-29
[12] Zhao B, Che C, Wang W, Li N and Cunningham B T 2021 Single-step, wash-free digital immunoassay for rapid quantitative analysis of serological antibody against SARS-CoV-2 by photonic resonator absorption microscopy *Talanta* **225** 122004
[13] Soler M, Estevez M C, Cardenosa-Rubio M, Astua A and Lechuga L M 2020 How nanophotonic label-free biosensors can contribute to rapid and massive diagnostics of respiratory virus infections: COVID-19 Case *ACS Sens.* **5** 2663–78
[14] Ravina D A, Mohan H, Prasad M and Pundir C S 2020 Detection methods for influenza A H1N1 virus with special reference to biosensors: a review *Biosci. Rep.* **40** BSR20193852
[15] Ribeiro B V, Cordeiro T A R, Freitas G R O, Ferreira L F and Franco D L 2020 Biosensors for the detection of respiratory viruses: a review *Talanta Open* **2** 100007
[16] Saylan Y and Denizli A 2018 Molecular fingerprints of hemoglobin on a nanofilm chip *Sensors* **18** 3016
[17] Rashed M Z, Kopechek J A, Priddy M C, Hamorsky K T, Palmer K E, Mittal N, Valdez J, Flynn J and Williams S J 2021 Rapid detection of SARS-CoV-2 antibodies using electrochemical impedance-based detector *Biosens. Bioelectron.* **171** 112709
[18] Battal D, Akgönüllü S, Yalcin M S, Yavuz H and Denizli A 2018 Molecularly imprinted polymer based quartz crystal microbalance sensor system for sensitive and label-free detection of synthetic cannabinoids in urine *Biosens. Bioelectron.* **111** 10–17
[19] Yang H, Qi D, Liu Z, Chandran B K, Wang T, Yu J and Chen X 2016 Soft thermal sensor with mechanical adaptability *Adv. Mater.* **28** 9175–81
[20] Wang Z *et al* 2016 Highly sensitive flexible magnetic sensor based on anisotropic magnetoresistance effect *Adv. Mater.* **28** 9370–7
[21] Teklemariam A D, Samaddar M, Alharbi M G, Al-Hindi R R and Bhunia A K 2020 Biosensor and molecular-based methods for the detection of human coronaviruses: a review *Mol. Cell. Probes* **54** 101662
[22] Saylan Y, Akgönüllü S, Yavuz H, Ünal S and Denizli A 2019 Molecularly imprinted polymer based sensors for medical applications *Sensors* **19** 1279
[23] Saylan Y, Yilmaz F, Özgür E, Derazshamshir A, Yavuz H and Denizli A 2017 Molecular imprinting of macromolecules for sensor applications *Sensors* **17** 898
[24] Liu L and Moore M D 2020 A survey of analytical techniques for noroviruses *Foods* **9** 318
[25] Dave P K, Rojas-Cessa R, Dong Z and Umpaichitra V 2021 Survey of saliva components and virus sensors for prevention of COVID-19 and infectious diseases *Biosensors* **11** 14
[26] Saylan Y, Yilmaz F, Özgür E, Derazshamshir A, Bereli N, Yavuz H and Denizli A 2018 Surface plasmon resonance sensors for medical diagnosis *Nanotechnology Characterization Tools for Biosensing and Medical Diagnosis* (Berlin: Springer) pp 425–58
[27] Antiochia R 2021 Developments in biosensors for CoV detection and future trends *Biosens. Bioelectron.* **173** 112777
[28] Abid S A, Muneer A A, Al-Kadmy I M S, Sattar A A, Beshbishy A M, El-Saber Batiha G and Hetta H F 2021 Biosensors as a future diagnostic approach for COVID-19 *Life Sci.* **273** 119117
[29] Saylan Y and Denizli A 2020 Virus detection using nanosensors *Nanosensors for Smart Cities* (Amsterdam: Elsevier) pp 501–11

[30] Saylan Y, Erdem Ö, Inci F and Denizli A 2020 Advances in biomimetic systems for molecular recognition and biosensing *Biomimetics* **5** 20
[31] Inci F, Saylan Y, Kojouri A M, Ogut M G, Denizli A and Demirci U 2020 A disposable microfluidic-integrated hand-held plasmonic platform for protein detection *Appl. Mater. Today* **18** 100478
[32] Kim S M, Lee T, Gil Y G, Kim G H, Park C, Jang H and Min J 2020 Fabrication of bioprobe self-assembled on Au–Te nanoworm structure for SERS biosensor *Materials* **13** 3234
[33] Regan B, Boyle F, O'Kennedy R and Collins D 2019 Evaluation of molecularly imprinted polymers for point-of-care testing for cardiovascular disease *Sensors* **19** 3485
[34] Wang J, Li J, Zeng C, Qu Q, Wang M, Qi W, Su R and He Z 2020 Sandwich-like sensor for the highly specific and reproducible detection of Rhodamine 6G on a surface-enhanced raman scattering platform *ACS Appl. Mater. Interfaces* **12** 4699–706
[35] He C, Ledezma U H, Gurnani P, Albelha T, Thurecht K J, Correia R, Morgan S P, Patel P, Alexander C and Korposh S 2020 Surface polymer imprinted optical fibre sensor for dose detection of dabrafenib *Analyst* **145** 4504–11
[36] Weber P, Riegger B R, Niedergall K, Tovar G E M, Bach M and Gauglitz G 2018 Nano-MIP based sensor for penicillin G: sensitive layer and analytical validation *Sensors Actuators* B **267** 26–33
[37] Mehrzad-Samarin M, Faridbod F, Dezfulu A S and Ganjali M R 2017 A novel metronidazole fluorescent nanosensor based on graphene quantum dots embedded silica molecularly imprinted polymer *Biosens. Bioelectron.* **92** 618–23
[38] Duan H, Li L, Wang X, Wang Y, Li J and Luo C 2015 A sensitive and selective chemiluminescence sensor for the determination of dopamine based on silanized magnetic graphene oxide-molecularly imprinted polymer *Spectrochim. Acta* A **139** 374–9
[39] Diltemiz S E and Uslu O 2015 A reflectometric interferometric nanosensor for sarcosine *Biotechnol. Prog.* **31** 55–61
[40] Li D Y, Qin Y P, Li H Y, He X W, Li W Y and Zhang Y K 2015 A 'turn-on' fluorescent receptor for detecting tyrosine phosphopeptide using the surface imprinting procedure and the epitope approach *Biosens. Bioelectron.* **66** 224–30
[41] Zhang X, Yang S, Jiang R, Sun L, Pang S and Luo A 2018 Fluorescent molecularly imprinted membranes as biosensor for the detection of target protein *Sensors Actuators* B **254** 1078–86
[42] Sharma S and Gupta B D 2018 An efficient and selective sensing of creatinine based on fiber optic SPR technique exploiting the advantages of molecular imprinting technique *Proc. SPIE* **10680** 1061801T
[43] Li D 2017 Molecular imprinting sensor based on quantum weak measurement *Biosens. Bioelectron.* **94** 328–34
[44] Liu S, Cheng R, Chen Y, Shi H and Zhao G 2018 A simple one-step pretreatment, highly sensitive and selective sensing of 17b-estradiol in environmental water samples using surface-enhanced Raman spectroscopy *Sensors Actuators* B **254** 1157–64
[45] Cavalcante F T T *et al* 2021 Designing of nanomaterials-based enzymatic biosensors: synthesis, properties, and applications *Electrochem.* **2** 149–84
[46] Dibekkaya H, Saylan Y, Yılmaz F, Derazshamshir A and Denizli A 2016 Surface plasmon resonance sensors for real-time detection of cyclic citrullinated peptide antibodies *J. Macromol. Sci.* A **53** 585–94

[47] Erdem Ö, Saylan Y, Cihangir N and Denizli A 2019 Molecularly imprinted nanoparticles based plasmonic sensors for real-time *Enterococcus faecalis* detection *Biosens. Bioelectron.* **126** 608–14

[48] Huang J, Tong J, Luo J, Zhu Y, Gu Y and Liu X 2018 Green synthesis of water-compatible fluorescent molecularly imprinted polymeric nanoparticles for efficient detection of paracetamol *ACS Sustain. Chem. Eng.* **6** 9760–70

[49] Primo E N, Kogan M J, Verdejo H E, Bollo S, Rubianes M D and Rivas G A 2018 Label-free graphene oxide-based surface plasmon resonance immunosensor for the quantification of galectin-3, a novel cardiac biomarker *ACS Appl. Mater. Interfaces* **10** 23501–08

[50] Inci F, Karaaslan M G, Mataji-Kojouri A, Shah P A, Saylan Y, Zeng Y, Avadhani A, Sinclair R, Lau D T Y and Demirci U 2020 Enhancing the nanoplasmonic signal by a nanoparticle sandwiching strategy to detect viruses *Appl. Mater. Today* **20** 100709

[51] Bodoki A E, Iacob B C and Bodoki E 2019 Perspectives of molecularly imprinted polymer-based drug delivery systems in cancer therapy *Polymers* **11** 2085

[52] Koca Esentürk M, Akgönüllü S, Yılmaz F and Denizli A 2019 Molecularly imprinted based surface plasmon resonance nanosensors for microalbumin detection *J. Biomater. Sci. Polym. Ed.* **30** 646–61

[53] Saylan Y and Denizli A 2019 Molecularly imprinted polymer-based microfluidic systems for point-of-care applications *Micromachines* **10** 766

[54] Li P, Lee G H, Kim S Y, Kwon S Y, Kim H R and Park S 2021 From diagnosis to treatment: recent advances in patient-friendly biosensors and ımplantable devices *ACS Nano* **5** 1960–2004

[55] Van Sau N, Ngo Q M, Phan T B, Tran N Q and Nguyen T T 2020 Optical biosensor using near infrared laser for enhancement of detection accuracy *J. Electron. Mater.* **49** 7420–6

[56] Kariithi H M, Welch C N, Ferreira H L, Pusch E A, Ateya L O, Binepal Y S, Apopo A A, Dulu T D, Afonso C L and Suarez D L 2020 Genetic characterization and pathogenesis of the first H9N2 low pathogenic avian influenza viruses isolated from chickens in Kenyan live bird markets *Infect. Genet. Evol.* **78** 104074

[57] Zhang N *et al* 2020 Recent advances in the detection of respiratory virus infection in humans *J. Med. Virol.* **92** 408–17

[58] Xu L, Liu J, Lu M, Yang D and Zheng X 2020 Liver injury during highly pathogenic human coronavirus infections *Liver Int.* **40** 998–1004

[59] Chen L and Liang J 2020 An overview of functional nanoparticles as novel emerging antiviral therapeutic agents *Mater. Sci. Eng.* C **112** 110924

[60] Qiu G, Gai Z, Tao Y, Schmitt J, Kullak-Ublick G A and Wang J 2020 Dual-functional plasmonic photothermal biosensors for highly accurate severe acute respiratory syndrome coronavirus 2 detection *ACS Nano* **14** 5268–77

[61] Moitra P, Alafeef M, Dighe K, Frieman M B and Pan D 2020 Selective naked-eye detection of SARS-CoV-2 mediated by N gene targeted antisense oligonucleotide capped plasmonic nanoparticles *ACS Nano* **14** 7617–27

[62] Seo G *et al* 2020 Rapid detection of COVID-19 causative virus (SARS-CoV-2) in human nasopharyngeal swab specimens using field-effect transistor based biosensor *ACS Nano* **14** 5135–42

[63] Liu H *et al* 2021 Development of a SERS-based lateral flow immunoassay for rapid and ultra-sensitive detection of anti-SARS-CoV-2 IgM/IgG in clinical samples *Sensors Actuators* B **329** 129196

[64] Huang L *et al* 2021 One-step rapid quantification of SARS-CoV-2 virus particles via low-cost nanoplasmonic sensors in generic microplate reader and point-of-care device *Biosens. Bioelectron.* **171** 112685

[65] Diao W, Tang M, Ding S, Li X, Cheng W, Mo F, Yan X, Ma H and Yan Y 2018 Highly sensitive surface plasmon resonance biosensor for the detection of HIV-related DNA based on dynamic and structural DNA nanodevices *Biosens. Bioelectron.* **100** 228–34

[66] Fu X, Cheng Z, Yu J, Choo P, Chen L and Choo J 2016 A SERS-based lateral flow assay biosensor for highly sensitive detection of HIV-1 DNA *Biosens. Bioelectron.* **78** 530–7

[67] Kurdekar A D, Chunduri L A A, Chelli S M, Haleyurgirisetty M K, Bulagonda E P, Zheng J, Hewlett I K and Kamisetti V 2017 Fluorescent silver nanoparticle based highly sensitive immunoassay for early detection of HIV infection *RSC Adv.* **7** 19863–77

[68] Luo L, Zhang F, Chen C and Cai C 2019 Visual simultaneous detection of hepatitis A and B viruses based on a multifunctional molecularly imprinted fluorescence sensor *Anal. Chem.* **91** 15748–56

[69] Sciuto E L, Petralia S, Calabrese G and Conoci S 2020 An integrated biosensor platform for extraction and detection of nucleic acids *Biotechnol. Bioeng.* **117** 1554–61

[70] Kashif M, Majeed M I, Hanif M A and Rehman A 2020 Surface enhanced Raman spectroscopy of the serum samples for the diagnosis of Hepatitis C and prediction of the viral loads *Spectrochim. Acta* A **242** 118729

[71] Seymour E, Lortlar Ünlü N, Carter E P, Connor J H and Ünlü M S 2021 Configurable digital virus counter on robust universal DNA chips *ACS Sens.* **6** 229–37

[72] Sharma P K, Kumar J S, Singh V V, Biswas U, Sarkar S S, Alam S I, Dash P K, Boopathi M, Ganesan K and Jain R 2020 Surface plasmon resonance sensing of Ebola virus: a biological threat *Anal. Bioanal. Chem.* **412** 4101–12

[73] Natesan M, Wu S W, Chen C I, Jensen S M R, Karlovac N, Dyas B K, Mudanyali O and Ulric R G 2019 A smartphone-based rapid telemonitoring system for Ebola and Marburg disease surveillance *ACS Sens.* **4** 61–8

[74] Draz M S *et al* 2018 Motion-based immunological detection of Zika virus using Pt-nanomotors and a cellphone *ACS Nano* **12** 5709–18

[75] Rong Z, Wang Q, Xiao N S, Wang J K, Xiao R and Wang S 2019 Smartphone-based fluorescent lateral flow immunoassay platform for highly sensitive point-of-care detection of Zika virus nonstructural protein 1 *Anal. Chim. Acta* **1055** 140–7

[76] Zhang Q, Zeininger L, Sung K J, Miller E A, Yoshinaga K, Sikes H D and Swager T M 2019 Emulsion agglutination assay for the detection of protein–protein interactions: an optical sensor for Zika virus *ACS Sens.* **4** 180–4

[77] Weerathunge P, Ramanathan R, Torok V A, Hodgson K, Xu Y, Goodacre R, Behera B K and Bansala V 2019 Ultrasensitive colorimetric detection of murine norovirus using NanoZyme aptasensor *Anal. Chem.* **91** 3270–6

[78] Nasrin F, Chowdhury A D, Takemura K, Lee J, Adegoke O, Deo V K, Abe F, Suzuki T and Park E Y 2018 Single-step detection of norovirus tuning localized surface plasmon resonance-induced optical signal between gold nanoparticles and quantum dots *Biosens. Bioelectron.* **122** 16–24

[79] Takemura K, Lee J, Suzuki T, Harad T, Abe F and Park E Y 2019 Ultrasensitive detection of norovirus using a magnetofluoroimmunoassay based on synergic properties of gold/magnetic nanoparticle hybrid nanocomposites and quantum dots *Sensors Actuators* B **296** 126672

[80] Chen H, Park H G, Choi N, Moon J I, Dang H, Das A, Lee S, Kim D G, Chen L and Choo J 2020 SERS imaging-based aptasensor for ultrasensitive and reproducible detection of influenza virus A *Biosens. Bioelectron.* **167** 112496

[81] Zhao X, Tsao Y C, Lee F J, Tsai W H, Wan C H, Chuang T L, Wu M S and Lin C W 2016 Optical fiber sensor based on surface plasmon resonance for rapid detection of avian influenza virus subtype H6: initial studies *J. Virol. Methods* **233** 15–22

[82] Lee J, Takemur K and Park E Y 2018 Plasmonic/magnetic graphene-based magnetofluoro-immunosensing platform for virus detection *Sensors Actuators* B **276** 254–61

[83] Peng X, Zhang Y, Lua D, Guo Y and Guo S 2019 Ultrathin Ti_3C_2 nanosheets based 'off-on' fluorescent nanoprobe for rapid and sensitive detection of HPV infection *Sensors Actuators* B **286** 222–9

[84] Guerreiro M R, Freitas D F, Alves P M and Coroadinha A S 2019 Detection and quantification of label-free infectious adenovirus using a switch-on cell-based fluorescent biosensor *ACS Sens.* **4** 1654–61

[85] Teengam P, Siangproh W, Tuantranont A, Vilaivan T, Chailapakul O and Hen C S 2017 Multiplex paper-based colorimetric DNA sensor using pyrrolidinyl peptide nucleic acid-induced AgNPs aggregation for detecting MERS-CoV, MTB, and HPV oligonucleotides *Anal. Chem.* **89** 5428–35

[86] Kim H, Park M, Hwang J, Kim J H, Chung D R, Lee K S and Kang M 2019 Development of label-free colorimetric assay for MERS-CoV using gold nanoparticles *ACS Sens.* **4** 1306–12

IOP Publishing

Biosensors for Virus Detection

Adil Denizli and Yeşeren Saylan

Chapter 3

Electrochemical biosensors for virus detection

Neslihan Idil and Bo Mattiasson

Viruses can be defined as very small and simple pathogenic agents. From the past to the present, viruses have caused global health emergencies to be declared. Based on these problems, therefore, there is a growing need and demand for rapid, reliable, real-time, sensitive, and selective detection methodologies using complex matrices. A few pathogenic viruses have caused much greater problems than others, and therefore, the focus has been on designing good analytical instruments and routines. The accurate sensing capability of electrochemical sensors represents a significant competitive advantage, which has gained great clinical attention for the detection of viruses and which has shown itself to be superior to traditional detection methods.

This chapter presents a review of electrochemical analytical biosensors and their monitoring applications. The operation of different devices is based on different electrochemical concepts. The sensitivities and convenience of use of different biosensors differ considerably, but together they offer a useful setup that is important in healthcare today.

3.1 Introduction

From the past to the present, infections have appeared in the water, food, air, and body; therefore, infectious agents are of significant interest, as they are causes of serious and life-threatening symptoms and even the death of living organisms [1, 2]. Viruses are, despite their size, extremely potent and constitute some of the most dangerous infective causative agents for human health in both industrialized and developing countries, also affecting the global economy [2, 3]. Following the notable emergence of viral outbreaks and their rapid spread, mortality and morbidity increase; as a consequence, such events become pandemics, and the scientific community leads the way in dealing with this challenge [2]. Following the emergence of influenza (H1N1) in 2009–2010 [4] and the appearance of Ebola virus in Africa in 2014 [5], the pandemic that now (2020–2021) strongly affects life in almost every country on Earth is the result of a viral presence named 'severe acute respiratory

 © IOP Publishing Ltd 2021

syndrome-related coronavirus 2 (SARS–CoV-2)'. SARS–CoV-2 originated in China at the end of 2019 and is the cause of (COVID-19) infection [4]. Diagnostic and antibody test kits have been applied intensively in clinical practice and alternative emerging trends that will serve the same purpose remain as priorities on the agenda [6]. Besides the coronavirus that is a severe threat to public and global health today, there are many other viruses, such as human immunodeficiency virus (HIV), hepatitis, Ebola, Zika, norovirus, and influenza that also may cause negative effects on life, such as hard-to-treat infections that are also difficult to detect [7]. The rapid and irreversible mutation of viruses manipulates the immune reactions of the host cell and prevents immune response mechanisms from functioning. In addition, viruses are adaptable to fast-changing environments and their particles can easily be released outside the host cells; incidentally, they cannot be treated with antibiotics due to the absence of enzyme systems. The reasons that the hard-to-treat infections caused by viruses have become almost impossible to cope with can be attributed to inadequate analysis methods and treatment strategies. This is because there is insufficient medical care that has feasible approaches for diagnosis and eventual treatment options. As a result of the additional advances in microfabrication technology, leading to improvements in molecular diagnostic methods, the microchip approach has come to the fore, and this technology gained momentum with the completion of the Human Genome Project.

Globally, it is necessary to discover viral sources by developing a promising strategy for threat inspection. The detection of viruses has now reached a flash point, and has become a very important daily activity in order to try to minimize the spread of the virus. It is worth noting that the identification and quantification of viruses play a significant role in detecting outbreaks and also in taking measures in clinical healthcare settings [3]. From this point of view, analytical devices such as biosensors provide an opportunity to operate with straightforward analytical platforms and cost-effective analytical devices, and improvements to the various sensing approaches are of great interest and a focus point in current research literature [8, 9]. An important part of these studies is to focus on the development of biosensors designed to monitor targets in real time without needless delay. The basic principle of biosensors can be explained in terms of the detection of a target analyte. In this context, the detector detects the reaction that occurs between the target molecule and the bioreceptor, that is to say, a physical or chemical binding reaction ultimately generates an alteration that can be transformed into a signal by a transducer, and measured by a detector. A biosensor should recognize the relevant virus and then initiate a signal indicating the detection of the virus that the sensor was built to analyze, and it should also be possible to measure the number of viruses present [10]. Intensive studies have taken place to develop biosensors that can operate rapidly and also with high sensitivity and selectivity [11–18]. The choice of the affinity reagent on the sensor chip is an important issue when setting up assays. Another decision is the choice of the type of sensor. Among different biosensors, one can identify, for example, optical sensors and electrochemical sensors, which include different components, such as the dominant electrochemical or optical transducers. This chapter is focused on the use of electrochemical biosensors for the detection of viruses.

In this chapter, we aim to address the recent developments and updates in the field of electrochemical sensing detection methodologies for viruses. For this purpose, we give a brief overview by describing the fundamentals and advantages of electrochemical sensing, including the significance of virus detection for global health. The applications of the electrochemical sensing groups developed for virus detection are broadly explained. Finally, challenges and future perspectives are discussed.

3.2 Virus infection and detection

3.2.1 Viruses

Viruses can make use of DNA or RNA as their genomic material and a protein capsid is found outside of the genome. Furthermore, a lipid envelope covers the capsid in several animal viruses. The infection process is driven by the absolute necessity of viral replication, and therefore viruses look for sensitive cells. The interaction between virus and host cell can be attributed to the receptor–antireceptor interaction. After this adsorption process, penetration takes place and the virus attached to the cell surface enters the host cell by inclusion in a very short time. Due to enzymatic reactions in endosomes, the capsid partially or completely disintegrates allowing the virus's nucleic acid to be released. Then, genome replication and translation take place and these processes differ according to the genomic characteristics of the virus. Newly synthesized viral proteins cover nucleic acid by virtue of their symmetry [4]. In the next step, the virus particle becomes infectious and capsid final rearrangements occur between proteins, the viral genome, and proteases. Subsequently, the mature viruses scatter from the cell they originated in [19].

3.2.2 Recognition elements for virus detection

Nucleic acids, proteins, particles of viruses, and also antibodies produced by individuals can be used as analytical targets for sensing [2]. When searching for the presence of viruses, one can register (i) whole virus particles, but also (ii) capsid proteins released from the virus's capsid (capsid proteins: e.g. VP1 of enterovirus 71 [20] and p 24 of HIV-1 [21]); (iii) furthermore, nucleic acids (DNA or RNA) originating from the virus and (iv) other targets can be analyzed to confirm that the virus has been present and caused an antibody immune response.

Several electrochemical sensing implementations for the detection of viruses are based on the application of bioreceptors including immobilized antibodies, antigens, and nucleic acids (figure 3.1). Immunodiagnostics has been the dominant analysis method, affording great potential, and antibodies raised against specific structures on the virus are also useful [22]. The outer surface antigens of viruses can be detected easily. Sensing tools include receptors or antibodies that can bind to these antigen structures; the reverse is also possible [4]. However, some aspects must be taken into account in order to establish electrochemical sensing systems. Along with the specific recognition ability and sensitive detection, robustness is one of the most essential properties of sensing platforms [9]. Bioreceptors that demonstrate a recognition ability for their specific antigens are mechanically weak under various circumstances, such as certain values of pH, ionic strength, and temperature. While the

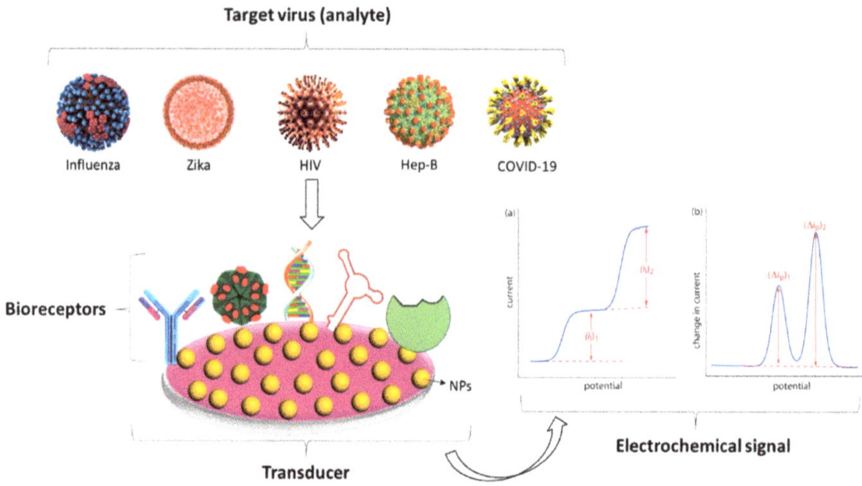

Figure 3.1. Potential electrochemical biosensor platforms for the detection of various pathogenic viruses. Reprinted from [25], Copyright 2020, with permission from Elsevier.

target molecule and bioreceptor interactions take place in the sensing stage, protection of the bioreceptor present on the synthesized matrix is required at the same time. In this context, it is possible to design systems with the desired properties by improving the biocompatibility and durability of sensors using innovative approaches to immobilization, conductive polymers, and nanomaterials [4].

Apart from these, the viral genome can be detected using sensors. If one wants to detect and/or analyze the presence of viral DNA/RNA, then complementary nucleic acid or its transcripted structures are used to perform hybridization [23]. In a previous study, single-stranded hepatitis B virus DNA was immobilized to a gold electrode. Given that 3'OH-DNA binds to carboxylate ester, and using electrochemical sensing, DNA hybridization could be easily followed with a label-free and high sensitive sensing strategy [24].

The latest findings in the field of electrochemical biosensors involve miniaturization of the sensor [26]. Furthermore, new affinity reagents have been introduced to replace the biomolecules that were used previously (antibodies, lectins, and sometimes enzymes). The new affinity reagents belong to the group of molecularly imprinted polymers (MIPs). In recent years, MIPs have become very popular when designing biosensors. Different means are used to produce MIPs. Their important characteristics are their high stability and the fact that MIPs are relatively easy to produce, as compared to the production of e.g. monoclonal antibodies. With sensors, it is attractive to reuse the affinity capturing structure, and, in that case, MIPs are superior to monoclonal antibodies and other biomolecules [8, 9]. A range of publications has described the use of electrochemical biosensors in which MIPs have been used as the affinity binder. An interesting development in recent years is the use of miniaturized MIPs.

MIPs have been prepared using an easy and cost-effective technology. Functional monomers and crosslinkers are copolymerized in the presence of the target. After the monomers are arranged around the target, it is allowed to perform polymerization on the selected matrix. The polymeric matrix includes three-dimensional capture sites that are produced with the shape and size of the target and chemical complementarity with it. Consequently, MIPs with high selectivity and affinity for the target molecules can recognize and bind/rebind such molecules [9, 27]. On the other hand, MIPs can be easily combined with electrochemical sensing approaches and have been applied in this way for the detection of viruses. An overview of the MIP preparation procedure, the signal enhancement process, and the sensing mechanism of electrochemical MIP-based sensors is given in figure 3.2.

The virus-based imprinting technique, which is identified as a state-of-art technology, holds great promise in the diagnosis of viral infections. The easy-to-apply approach is surface imprinting. The geometry and structure of the surface, e.g. capsid proteins, enable the monitoring of the interaction between the imprinted site and the virus [29].

When a virus is infecting a patient, then there may be a spread of capsid proteins. In healthcare, there might be situations in which staff members are injured by e.g. the needle of a syringe that has been used to treat an infected patient. Such a mistake might result in the transfer of viruses from the blood of the patient to the healthcare professional [30]. In the case of patients with HIV, the professional may then have a tough period, since they will not know if they have been infected or not. The normal process is to wait for up to three months and then evaluate whether the production of antibodies against HIV has taken place. If no antibodies are formed then one can conclude that there was no transfer of virus particles in the accident.

An electrochemical biosensor based on imprinted multiwalled carbon nanotubes (MWCNTs) was developed to detect HIV-p24. A detailed procedure diagram for the fabrication of the sensor is given in figure 3.3: (i) MWCNTs were deposited on the electrode; (ii) CS coating was performed on the electrode and an amine layer was produced; (iii) glutaraldehyde activation was carried out to introduce aldehyde groups; (iv) HIV-p24 was applied and bound via an imine bond between the amino

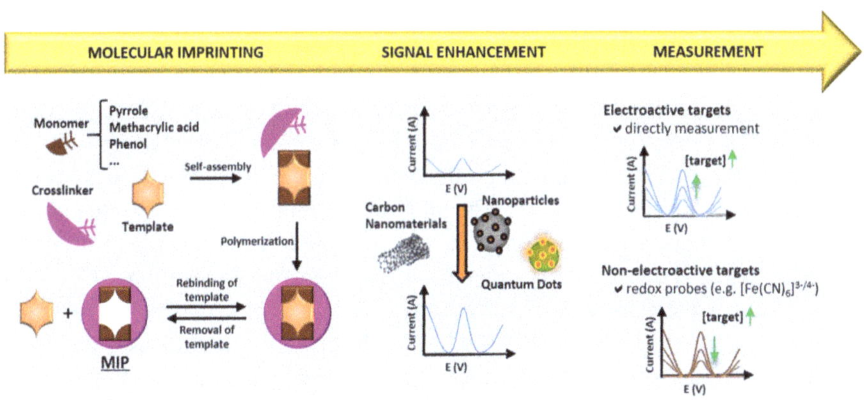

Figure 3.2. Summary of the MIP preparation procedure, signal enhancement process, and the sensing mechanism of electrochemical MIP-based sensors. Reprinted from [28], Copyright 2021, with permission from Elsevier.

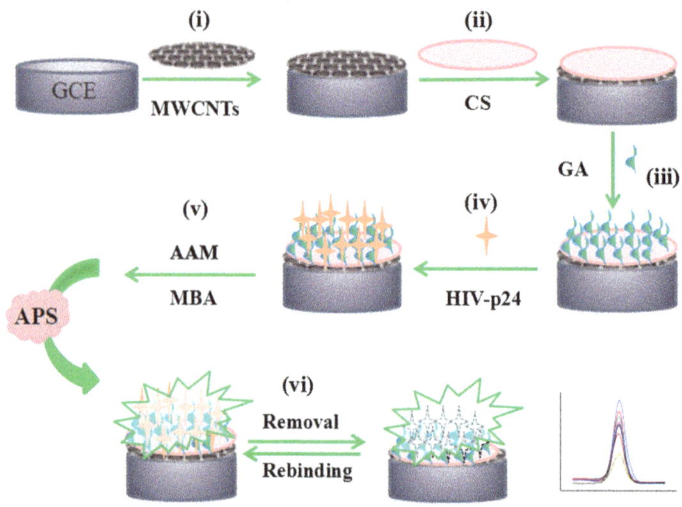

Figure 3.3. Detailed procedure diagram for the fabrication of the HIV-p24-MIP/MWCNT/glassy carbon electrode (GCE) sensor. Reprinted from [31], Copyright 2021, with permission from Elsevier.

group of the protein and the aldehyde groups of glutaraldehyde; (v) polymerization was performed using acrylamide as a functional monomer and ammonium persulfate as an initiator and cross-linker. The availability of the proposed sensor was verified by enzyme-linked immunosorbent assay (ELISA). It was pointed out that the most desirable feature of the sensor was that specific and sensitive capturing was successfully achieved for HIV-p24 with a sensitivity of 0.083 pg cm^{-3} [31].

In another study, a surface imprinting-based electrochemical sensor was developed using a graphene oxide composite to detect the Zika virus (figure 3.4). The analytical performance of the proposed system in buffer indicated that the limit of detection (LOD) was 2×10^{-2} PFU mL^{-1} when the dengue virus was present as a competitor. Ten percent and one percent diluted serum samples were used to investigate the effectiveness of the sensing in real samples and the LODs were reported to be 2×10^{-3} and 5×10^{-2} plaque-forming units (PFUs) mL^{-1}, respectively. These results endorse the use of imprinting strategy-based sensing in clinical healthcare settings [32].

However, one may try to develop a quicker analysis, in which case, it might be attractive to try to detect the presence of capsid proteins. For such an assay to be successful, it must be very sensitive. In collaboration with the Food and Drug Administration (FDA) in the USA, the staff of CapSenze AB in Sweden utilized a capacitive sensor with high sensitivity. It was possible to obtain a yes or no response within one week. The sensitivity of the capsid protein assay was 10^{-17} M—a sensitivity that at the time was outstanding.

A range of viruses has been identified as pathogenic organisms that have caused severe illness among infected individuals. Virus-caused diseases are very difficult to handle due to the lack of potent drugs that are active against the viruses. Much research attention has been paid to the development of the current diagnostic tools and to speeding up detection. As an indication of the seriousness of this topic,

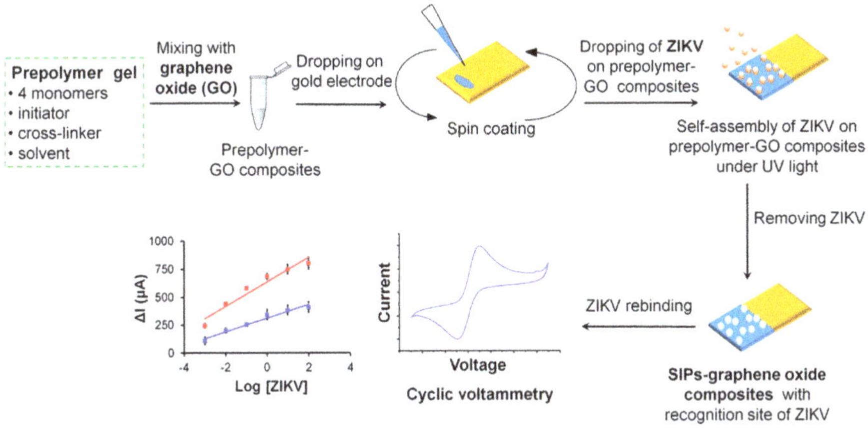

Figure 3.4. Scheme depicting the preparation of surface imprinted (SIP)–graphene oxide composites on a gold surface to produce a Zika virus biosensor. When the Zika virus was added to the composites, there was a chance that some particulate matter would be captured on the copolymer sites. At that time, the copolymers were partially polymerized and a self-assembly process occurred due to the presence of some prepolymer around the virus. After the polymerization process had finished, the viruses were removed and the recognition sites became cavities. Reprinted with permission from [32]. Copyright 2019 American Chemical Society.

the 2020 Nobel prize for medicine was awarded to the researchers who identified the hepatitis C virus and also the means to treat the disease [33]. Viruses that cause disease include coronaviruses, HIV, hepatitis B and C, Ebola, Zika, norovirus, influenza, and rabies. Apart from the viruses that cause human disease, a range of other viruses causes problems in e.g. the production of food/feed, for example, avian influenza virus [33] and the virus responsible for white spot syndrome in shrimp cultivation water [34].

A relatively small group of viruses causes severe illness, and there have been strong efforts to design suitable sensors for these virus groups.

Coronavirus

Coronaviruses (CoVs) are round, enveloped, and single-stranded RNA viruses and are known to be dangerous agents of acute and chronic respiratory and gastro-intestinal infections. Severe acute respiratory syndrome (SARS) and Middle East respiratory syndrome (MERS) can occur due to the infection of epithelial cells. Transmission can be attributed to aerosol contact. Although SARS-CoV-2 is usually transmitted by droplets (sneezing, coughing), it has also been proven that people can also be infected by touching a surface or object with the virus on it and then touching their own eyes, nose, or mouth. For the diagnosis, the reverse transcription-polymerase chain reaction (RT-PCR) is widely applied [35]. The pathogenesis of the infection first starts with attachment to the host cell, following which, the 'S' protein interacts with the receptor. Severe acute respiratory syndrome coronavirus 2 (SARS-CoV-2) binds to angiotensin-converting enzyme 2 (ACE2) via its spike protein. The expression of the receptor takes place in the lungs, leading to respiratory problems [4]. In addition, proteins, structural properties, and enzymatic reactions have a crucial role in the detection of the virus based on the host cell's immune response [36]. In this regard, clinical symptoms can be observed, such as fever, cough, trouble breathing, and diarrhea. If a complication occurs, systemic problems arise in vital organs, including the heart and liver. Over time, oxygen saturation decreases, and ventilator support is required [37]. The seriousness of the infection increases in elderly patients.

In the last two years, over 32 million patients have suffered from SARS-CoV-2 and millions of fatal cases have been recorded. Widespread vaccination and vaccines currently under development will contribute to controlling pandemics. The research field concerned with the development of faster, easier, low-cost, and precise testing for COVID-19 has become important and is growing day by day. Blood samples from individuals with suspected coronavirus are subjected to RT-PCR, ELISA, IgG, and IgM antibody tests [16]. As a result, new measurement methods are of great interest in the literature and alternative approaches to produce different clinical kits and/or devices based on this strategy have gained attention.

Yakoh *et al* designed a paper-based electrochemical sensor to capture immunoglobulin against SARS-CoV-2 (figure 3.5). If specific antibodies for the virus are present, the redox change stops and the sensor response is declined. It has been indicated that the developed sensing tool can be applied to blood samples.

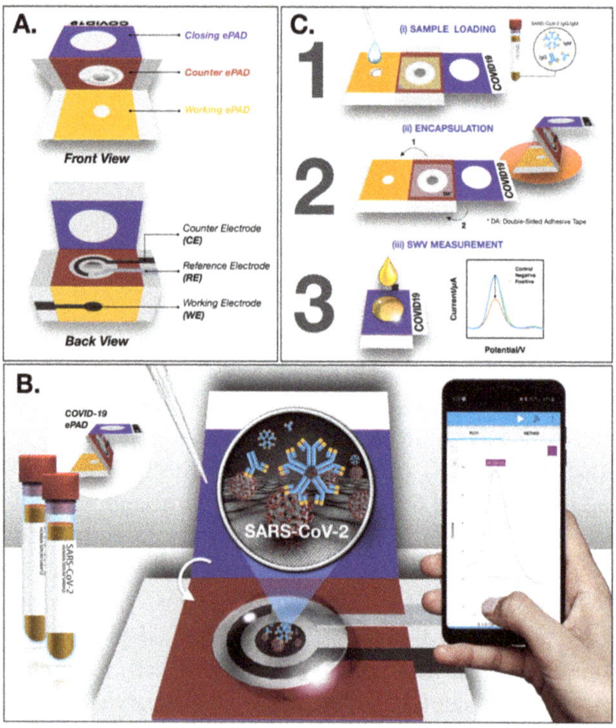

Figure 3.5. Schematic illustration of (A) the device components, (B) the detection principle, and (C) the detection procedure of the COVID-19 electrochemical PAD (ePAD). Reprinted from [38], Copyright 2019, with permission from Elsevier.

This electrochemical sensor has been proven effective using real clinical sera from patients with satisfactory results. One may also use the opposite approach: instead of analyzing antibodies, antigen detection may be possible [38].

Quantitative reverse transcription PCR (qRT-PCR) has gone one step further, but has some limitations, such as the expense of the devices and the necessity for a long analysis period. Products based on isothermal rolling circle amplification have been hybridized with probes modified with redox labels. The signals of the proposed single-step electrochemical technology were compared with the cycles of qRT-PCR; 100% verification was obtained, and therefore, this promising approach allows real-time detection to be considered [39].

The use of graphene materials for virus detection allows the development of point-of-care (POC) systems. Its low-cost production and fast and sensitive sensing make graphene a front-runner and the application of graphene heralds a new era due to its unique mechanical and electrochemical properties.

Alafeef *et al* [40] developed a rapid (<5 min) responsive, applicable, cost-effective, and quantitative paper-based electrochemical sensor for SARS-CoV-2. Their system included two main components: an electric signal measurement platform and probes used to detect viral RNA. In the first step, a conductive film layer was formed by coating filter papers with graphene. A gold electrode was placed onto the graphene as a contact surface for electrical readout. Both gold and graphene have high sensitivity and conductivity. Electron transfer accelerated by the gold nanoparticles propagates through the detection platform, causing an increase in the output signal, and thus confirming the presence of the virus.

Human immunodeficiency virus (HIV)

HIV is the cause of acquired immunodeficiency syndrome (AIDS) which is a pandemic infection [41]. HIV was first recognized at the beginning of the 1980s and its high mortality rate has created global concern. After a while, retroviral treatment allowed it to be controlled as a chronic disease, and since then, a decrease in AIDS-related deaths has been reported from year to year [42]. In diagnosis, ELISA, enzyme immunoassay, and western blotting were generally preferred. The most conspicuous feature of HIV is that the symptoms appear in due course after the HIV infection. The destroyed immune system fails to retain its protection ability since CD4-T lymphocytes are affected, HIV spreads throughout the vascular system, and blood-borne infections become inevitable [41].

DNA sequencing stands out as a valuable and significant methodology that creates a route for diagnosis. In a previous study, epoxy graphite was functionalized with graphene oxide to detect HIV-1. The performances of composite electrodes were tested by voltammetry and the oxidation signal of guanine of oligonucleotide was detected [43].

Yeter *et al* [44] fabricated a DNA-based electrochemical sensor with the help of carbonized glass fiber–coal tar pitch (GF–CTP) electrodes to detect the HIV-1 gene. GF–CTP electrodes were functionalized with thiol what and; gold nanoparticles were preferred for this platform and self-assembled to obtain good conductivity. A probe was attached to the resultant electrode and electrochemical impedance spectroscopy (EIS) was then used to carry out analytical experiments. The LOD was found to be 13 fM, and therefore, it was confirmed that carbon-based materials are promising tools for integration into sensing systems.

Hepatitis

The hepatitis B virus is one of the most common viral agents that affects the human metabolic system. It has been reported that 20% or more of infected patients could face liver malignancy, and cirrhosis can occur in later stages of the infection. In diagnosis, ELISA is frequently applied to detect hepatitis B surface antigen (HBsAg), hepatitis B surface antibody (anti-HBs), hepatitis B e-antigen (HBeAg), hepatitis B e-antibody (anti-HBe), and total hepatitis B core antibody (anti-HBc). Apart from these, PCR-based methods have also been used [17].

Hepatitis E virus infection is known to be the cause of acute hepatitis and it can be emerge as a result of enteric contagion. The main genotypes are endemic, and have become important public health threats. In addition, millions of people are infected every year, and fulminant hepatitis can occur in pregnancy. In diagnosis, anti-HEV antibodies and the RNA of the virus are detected by enzyme immunoassay and RT-PCR [3].

Ebola

The Ebola virus has a negative-strand RNA and causes hemorrhagic fevers with a mortality rate of 44%–90%; however, the infection is preventable by early detection [4]. Body fluids are the most common means of transmission. In diagnosis, RT-PCR is proven to detect Ebola and is routinely used; however, in the early stages, test results may be unreliable. In addition, immunoglobulins can sometimes be detectable, but in some cases, they are not detectable due to their absence. In this respect, sensing devices have even greater importance as high sensitivity diagnostic tools.

In research that aimed to detect the Ebola virus, DNA-based electrochemical sensing was performed by screen-printed electrodes (SPEs) and probes (figure 3.6). A thiolated DNA capture probe sequence was bound to the SPEs and allowed to hybridize with biotinylated DNA. Streptavidin-alkaline phosphatase was applied to form a biotin-streptavidin conjugation and the conversion of 4-aminophenyl to 4-aminophenol was then detectable. Sensing experiments were carried out using EIS using quantities of between 10–75 nM and the LOD was found to be 4.7 nM [45].

Zika

Zika virus is known to be the cause of an infectious disease transmitted by mosquitoes, which was first described in 1947 in the Rhesus monkey in Uganda [46]. Zika virus is known to be the reason for problems in the skin and eyes, malaise, muscle problems, weakness, headache, or mild fever. Infections did not originally attract much interest; however, the spread of the virus to many countries has made the situation serious, and the number of individuals suffering from Zika virus infection has been reported to be approximately 3–4 million. Increasing outbreaks have been recorded and the capability of the containment applications used for diagnosis in affected countries is creating a requirement for new technologies [32].

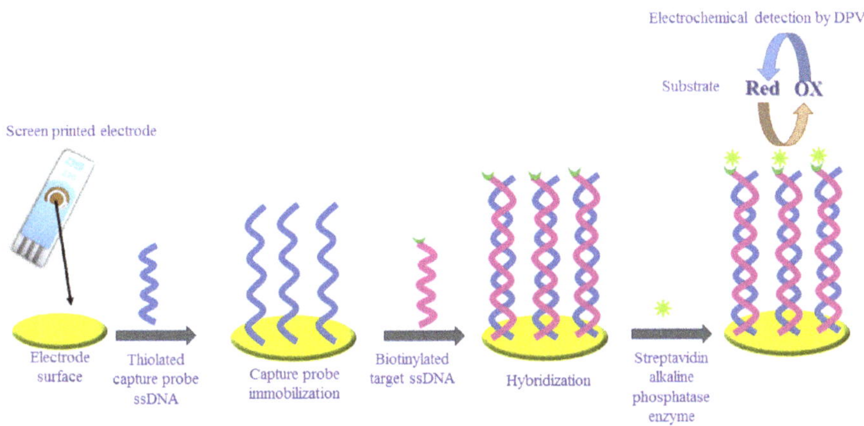

Figure 3.6. Schematic representation of different steps used in the fabrication of an electrochemical DNA biosensor. Reprinted from [45], Copyright 2018, with permission from Elsevier.

Kaushik *et al* researched and developed an immunosensor for Zika virus based on the immobilization of Zika virus-specific envelope protein antibody on an interdigitated gold microelectrode. The mechanism worked with no requirement for labeling and the Zika virus envelope protein was bound specifically to the immobilized antibodies on the gold microelectrode. An increase in the charge-carrying resistance originating from the decreasing rate of charge transfer was measured by EIS [47].

Influenza

Influenza viruses have frequently been seen worldwide over the years. They constantly change due to mutation; they are quite different from the virus seen in the previous year, and this situation leads to illness in the following years. The infections caused by these viruses have resulted in many socioeconomic collapses. The causative agents of influenza are the influenza A and B viruses. The most contagious is influenza A, which causes widespread disease in humans and animals such as pigs, birds, and horses, while influenza B only causes disease in humans. There are also subtypes of the influenza A virus that only cause disease in animals. The most well-known example of this is bird flu. In recent years, it has been shown that this virus can also be transmitted to humans. Although such transmissions represent a small number of cases, there are concerns that large outbreaks will occur [48].

Influenza is spread by coughing and sneezing, which scatters droplets carrying live viruses that cause illness. These droplets can remain in the air for several hours and transmit the disease to humans. For this reason, people in crowded environments, such as schools and workplaces, are especially at risk. In addition, virus-containing particles that are transmitted to the hand of a sick person and from there to objects, such as door handles, telephones, and so on, can cause the disease when another person comes into contact with them and then moves their hand to their mouth, nose, or eyes [49].

The recommended method for preventing influenza infection is vaccination. The World Health Organization closely monitors changes in the virus and makes annual recommendations for the vaccine composition. Every year, the vaccine content is prepared by considering the recommendations of the World Health Organization [50]. There is a necessity for an ongoing upgrade to obtain more capable systems and to cope with increasing challenges in terms of virus detection.

A previous study conducted for the detection of influenza A aimed to use electrochemical sensing. For this purpose, anti-H7N1 obtained from a vaccinated rabbit was bound to a modified electrode surface. Sensing experiments were performed using EIS, and the LOD was reported to be 5 μg mL^{-1} [51]. Another study aimed to fabricate genomic sensors using oligonucleotides specific to the H5N1 avian influenza virus A. A redox-active monolayer on a gold electrode surface was formed by a 'click' reaction between an amine and an epoxide group and the analytical performance showed that the LODs of a 280-mer RNA1 transcript and a 20-mer DNA target were 73 and 0.87 pM, respectively [15].

To give an example of the use of an immunomagnetic application and gold catalysis, influenza A virus H9N2 was detected by an chronoamperometric sensor (figure 3.7). Anti-matrix protein 2 antibodies bound to iron magnetic nanoparticles were used to obtain viruses in the sample. Fetuin A bound to gold nanoparticles, revealing its interaction with hemagglutinin. The target virus attached to these two receptors was obtained and treated with an acidic medium to get gold nanoparticles deposited on a screen-printed carbon electrode. The gold nanoparticles enabled the monitoring of the reduction of hydrogen ions. It was concluded that the proposed system was promising for the specific recognition of the target virus at a titer of less than 16 hemagglutinin units [52].

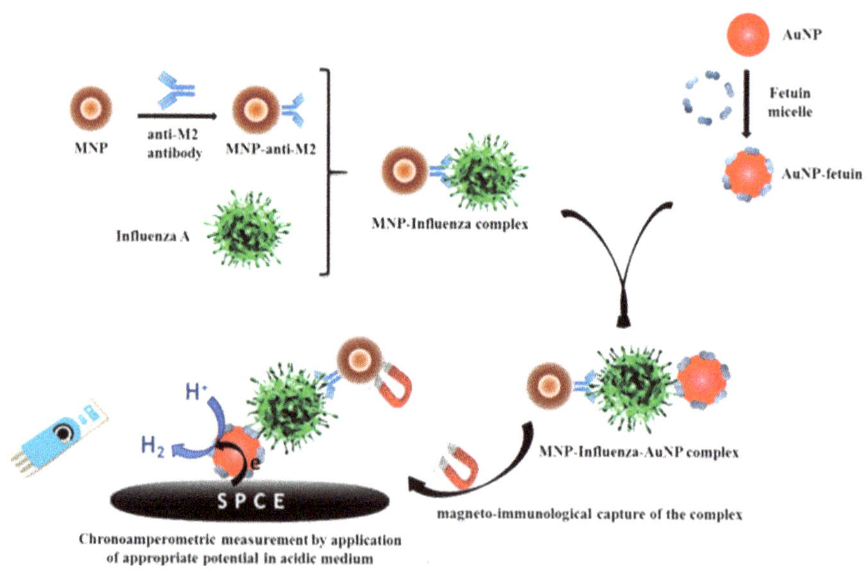

Figure 3.7. Schematic illustration of the strategy used to develop the gold nanoparticle-based chronoamperometric magneto-immunosensor for influenza virus. Reprinted from [52], Copyright 2018, with permission from Elsevier.

3.3 Electrochemical sensors

3.3.1 Principle of electrochemical sensors

Electrochemical sensors are based on chips with affinity ligands. When a target (which could be the target molecule) binds to a sensor, an electric signal is created which is used to analyze the concentration of the target substance. Electrochemical sensors are devices that include two main components; one is the recognition part (enzymes, antibodies, proteins, and molecularly imprinted polymers, etc.), while the other is a physicochemical transducer. The transducer can transform electrochemical information and produce a signal that is proportional to the analyte concentration. Working, reference, and counter electrodes are useful and preferred to obtain instrumental signals from transducer parts [27]. Reference electrodes can be created from Ag/AgCl and placed to maintain a stable potential. The sensing electrodes can be referred to as the transduction layer, and they have a function in the process of the reaction. In terms of virus detection, the electrode can be modified with viral proteins, a probe complementary to the viral genome, or antibodies [4]. Counter electrodes can be identified as contact materials that provide potential to the working electrode. The type of electrode preferred for use has a key role in the effectiveness of electrochemical sensors. In the research literature, carbon or metal electrodes, such as gold, platinum, and silicon (oxide) have been used for electrochemical sensing and have an extensive publication history. Reactions stimulate signals in the form of changes to conductance, impedance, or detectable current [53]. In this respect, potentiometric-, amperometric-, cyclic voltammetry-, capacitance-, and impedimetric-based sensing are the main types of this sensing technology.

3.3.2 Advantages of electrochemical sensors

Electrochemical sensors appear to be one step ahead in comparison to other sensing systems, due to their ability to provide fast, sensitive, and selective recognition with low limits of detection, with the additional advantages of portability and easy miniaturization. In addition, the use of a small quantity of chemicals and a low sample size are other favorable features [27]. In the last decade, electrochemical sensing has received great attention due to its application to the recognition of glucose, heavy metal ions, phenolic compounds, cancer biomarkers, nucleic acids, cells, microbial cells, and viruses. Electrochemical measurements can be applied to detect other compounds or cells via newly developed sensing strategies, enabling analytical sensitivity to be enhanced. In addition, electrochemical nanosensors can easily be integrated into point-of-care analysis instruments and offer the possibility of on-site detection and easy clinical diagnosis systems. In the light of these prominent benefits, electrochemical sensors can be considered superior for both diagnostics and the monitoring of infections and encouraging tools for the transformative alteration of clinical applications [53].

In recent years, nanomaterial-based approaches have attracted great interest and contributed to the development of viral electrochemical sensors. Nanomaterials can also be easily integrated into electrochemical sensing systems to obtain their advantage of increased surface area; the resulting sensing platforms are called nanosensors. The other opportunities offered by nanosensors are improved cost-effective, efficient, miniaturized, and portable systems with high-throughput screening along with low limits of detection. Apart from these benefits, nanomaterials can be simply and rapidly immobilized to the surface of interest, and using nanomaterials in sensing systems provides highly sensitive and selective detection [54]. These materials also have a crucial role in the development of high-performance analytical sensing platforms with the use of cutting-edge elements offering conductivity.

GCEs are reported to be one of the commonly applied electrodes. On the other hand, SPEs have frequently been used to enhance analytical performance due to their ability to create appropriate shapes, dimensions, and conductivity, since they can easily be modified with nanomaterials. The other significant reason to prefer SPEs is that only a small sample is required due to their small size and therefore, portable SPE-based sensing strategies are of great interest. Both GCEs and SPEs enable the establishment of attractive MIP-based sensors [55]. Another choice is to use graphene, which provides a high surface-to-volume ratio, high conductivity, and fast electron transfer. Carbon nanotubes (CNTs) and MWCNTs are widely applied and combined with electrochemical sensing platforms, since they are capable of electrical signal production [56]. Ma *et al* [30] established a sensing platform for HIV using electrochemical measurements with a multiwalled carbon nanotube modified glass carbon electrode. Huang *et al* [57] detected human papillomavirus using a recognition probe attached to a GCE modified with graphene/Au nanorod/polythionine.

A previous study was able to detect influenza H1N1, H5N1, and H7N9 viruses using ZnO nanorods interacting with polydimethylsiloxane. As a consequence, an LOD of 1 pg mL^{-1} was obtained with immobilized antibodies. The experimental range that offered good linearity was reported to be 1–10 ng mL^{-1}. In addition, the proposed platform was able to recognize each specific virus [58].

Conventional approaches based on cultivation, serological tests, the polymerase chain reaction (PCR), ELISA, and blotting have been widely used for the detection of these analytical targets. These methodologies are reported to be limited by their time-consuming and labor-intensive steps, the difficulty of obtaining pure cultures of viruses, the requirement for educated personnel and high-priced laboratory equipment, and the difficulty of transferring test materials to be analyzed in competent laboratories. Sensing approaches have been introduced as attractive alternatives to the conventional methodologies mentioned above and provide cost-effective, sensitive, fast-response, miniaturized, and portable systems [16].

3.3.3 Importance of electrochemical biosensors in medical diagnosis

Affinity-based analyses have been commonly used in medical diagnostics. The initial concept for analyses was to operate with labeled reagents, which simplified the evaluation of analyses [59]. The introduction of labeled antigens when evaluating binding to antibodies represented an important step towards more selective assays and also improved sensitivity. The use of labels was originally connected with practical problems. Radioisotopes were used as labels [60]. The half-life of some of the isotopes was short and limited the storage and use of those assays. Later, the radioisotopes were replaced, since it was clear that the handling of radioisotopes should be avoided. Thus, the radioisotopes were replaced by the use of fluorescent labels, enzymes, etc; ELISA [61] was the most used when, eventually, developments moved toward skipping the labels and instead registering when target molecules were bound directly to a sensor surface.

Strategies applied to functionalize the sensor surface allow the synthesis of new sensor-surface materials that offer high performance. However, there is still difficulty in developing sensor platforms that can detect with high selectivity and sensitivity and that also exhibit high reproducibility. In addition, the complex environments of naturally sourced samples present a great obstacle, in terms of interaction between target molecules and sensor surfaces. In this context, critical issues still remain in electrochemical biosensors that need practical solutions [6, 62].

3.3.4 Capacitive sensors

In a typical capacitive biosensor, the biological recognition element is properly attached to the electrode surface by the process of immobilization. Furthermore, the covering layer of the electrode must be tight to prevent ions from the bulk from coming into contact with the electrode surface. When the electrode is brought into the electrolyte solution, the orientations of ions are different from those of the case before the electrode surface was modified by placing tight layers over the electrode and also immobilizing reagents on the covering layers. Here, the electrical double layer (EDL) is forced away from the electrode surface. Bergdahl *et al* [6] described a representative capacitive biosensor electrode, in which a self-assembled monolayer (SAM) was used for the immobilization of the biological sensing element. The capacitance at an electrode/solution interface was then described as being built up of several capacitors in series. The thicker the covering layer (the first capacitor and each subsequent layer constitute a new capacitor) the thicker the spacers used for immobilization, the affinity reagent that captures the target, and also the target.

It is worth mentioning that the lowest capacitance dominates the total capacitance. As such, it is very important to design the surface so that the capacitance of the insulating layer is as high as possible, otherwise, it may dominate, and the capacitance due to the binding of the analyte may not be detected. The equation is written based on the assumption that the surface of the electrode is completely covered by the recognition element and that all the influences of pinholes are negligible. Otherwise, the capacitance is in parallel with the resistors, resulting in a decrease in sensitivity due to current leakage. This may also cause a conductive solution to short-circuit the capacitances of the layers on the electrode [8].

The CapSense team and co-workers produced a capacitive sensor for the detection of p24, a capsid protein of HIV-1. A time course diagram showing the decreases in capacitance after the injection of standard p24 is presented in figure 3.8.

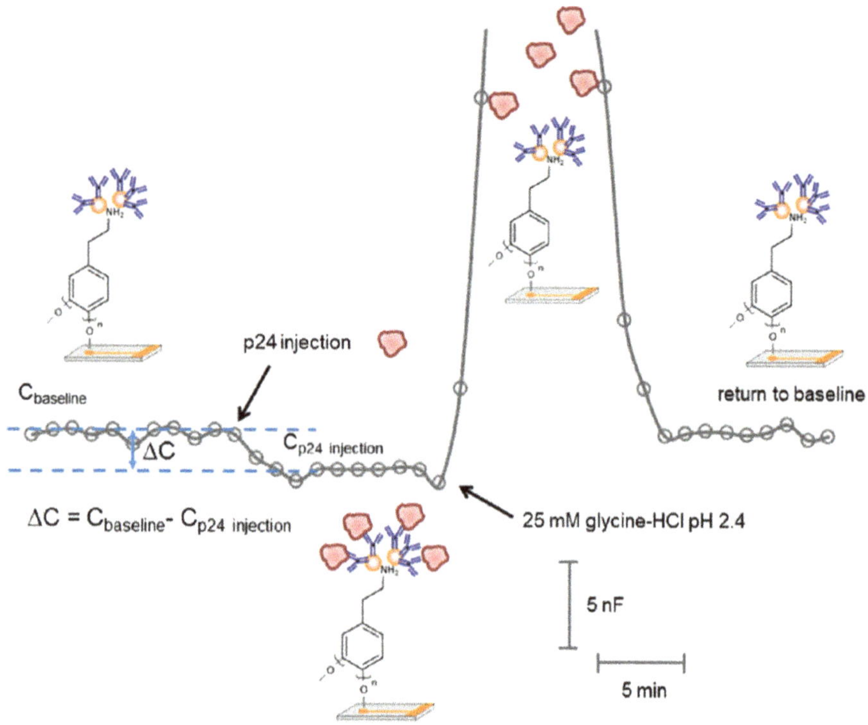

Figure 3.8. Time course diagram showing the decreases in capacitance after the injection of standard p24. The average capacitance levels, $C_{baseline}$ and C_{p24} injection, represent the levels used for the calculation of the capacitance change (ΔC). After the sample was injected, the surface was subsequently regenerated using 25 mM glycine–HCl, pH 2.4. Reprinted with permission from [21]. Copyright 2010 American Chemical Society.

The effectiveness of the proposed sensor was demonstrated by its ability to detect sub-attogram per milliliter concentrations of p24 [21].

To introduce selectivity to capacitive biosensors, different affinity reagents are immobilized on the electrode surface. However, as mentioned above, there is a requirement for a tight layer over the electrode material. This can be achieved by using self-assembling reagents (when a gold surface is to be covered, thiol reagents are preferred, since they form tight layers. [63]. Another strategy has been to use the electrocatalyzed polymerization of tyramine and thereby to form a polytyramine film. The different layers applied need to contain some reactive groups that can be used to immobilize molecules with an affinity for the target that is going to be quantified. Impedance is somewhat similar to capacitance. Both technologies have been used successfully [64, 65].

3.3.5 Impedimetric sensors

Impedance immunosensors are used to simultaneously detect the change in impedance at the interface or charge transfer resistance of an electroactive probe dissolved in solution before and after the target molecule is bound.

EIS has been introduced as an applicable and effective method because inappreciable changes that emerge from the interaction of the solution with the electrode can easily be monitored in real time without any requirement for labeling. Signal measurements were successfully performed, as confirmed by their ability to differentiate the binding process. On the other hand, EIS is commonly applied for the characterization of electrodes [25, 66]. To give an example, the hepatitis B virus was detected using a gold-binding polypeptide-attached antibody with an affinity for hepatitis B virus antigen and gold electrodes with a label-free methodology and it was suggested that the results obtained will introduce the rapid recognition of viruses to POC sensing [67].

A study aimed to detect noroviruses, which are the causative agents of acute foodborne gastroenteritis. A schematic illustration of the detection of norovirus using a novel peptide-coated electrochemical biosensor is shown in figure 3.9. Eight peptides were deposited on gold electrodes and the norovirus recognition abilities of the peptides were tested. As a consequence, it was confirmed that NoroBP peptide has good affinity for norovirus; there was a decrease in signal with an increasing concentration of Norovirus, with an LOD value of 1.7 copies mL^{-1}, which is three times lower than that of other reported methods [68].

3.3.6 Piezoelectric sensors

A piezoelectric sensor is based on an oscillating crystal. This concept has frequently been used in many applications, but has recently come to the fore for the construction of biosensors. A small quartz crystal will have a defined oscillation when exposed to an electric field. If material is deposited on the crystal surface, the oscillation frequency decreases. At the starting point of the construction of a

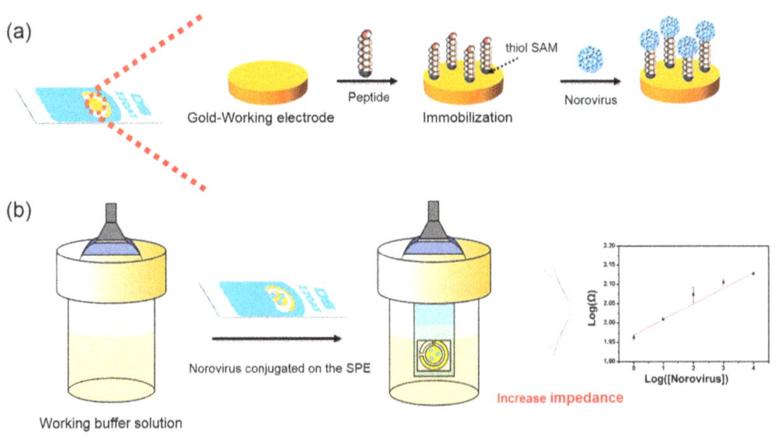

Figure 3.9. Schematic illustration of the detection of norovirus using a novel peptide-coated electrochemical biosensor. (a) The novel peptides were immobilized as SAMs on the Au working electrode. (b) Using a working buffer solution, which can be used for oxidation and reduction, the affinity strength of norovirus dropped on the Au working electrode was then measured using an EIS analysis. Reprinted from [68], Copyright 2019, with permission from Elsevier.

piezoelectric biosensor, an affinity binder is placed on the surface of the crystal. When a molecule binds to the affinity group, a change in the oscillation frequency takes place. The weight of the molecule that binds has an important effect on the amplitude of the piezoelectric sensor: the larger the molecule, the more pronounced the response. Piezoelectric sensors are thus very useful for the analysis of macromolecules or even particulate material. Assays of minor chemicals are somewhat more complicated [69–71].

In a previous study, influenza A was successfully recognized by piezoelectric discs functionalized using a sialyl glycopolymer receptor structure. Virus capture was monitored via a radial-mode resonance frequency shift. It was demonstrated that transducers with modified layers provide sensitive recognition with strong regeneration properties. The proposed label-free detection was performed at up to 10^5 virus mL^{-1} [69].

As an example of research into the ongoing pandemic, SARS-CoV was detected using a piezoelectric immunosensor. In this method, horse polyclonal antibody was attached to a piezoelectric matrix; antigenic determinant was then investigated in the sample. If the crystal surface with affinity reagents can catch its specific antigen, a frequency shift occurs due to mass alteration; linearity was found to occur at concentrations of 0.6–4.0 $\mu g\ mL^{-1}$ [70].

The blood-borne viruses of greatest concern are the human immunodeficiency viruses (HIV-1 and HIV-2). Bisoffi *et al* [71] developed a sensing system that relied on piezoelectric materials modified with antibodies for HIV-1 and HIV-2. Sensing experiments were carried out using human serum, and even competitors were found in complex media. The POC sensing platform created provides exact results and can be recommended as an appropriate testing tool [71].

3.4 Phage display technology – an interesting possibility for biosensors

As well as the use of electrochemical biosensors for the detection of viruses or biomolecules derived from them, another option is to combine viruses/bacteriophages with biosensors. A new technology to create affinity reagents has been presented and called 'phage display technology'. In this situation, the phage has been manipulated such that it produces specific proteins that are placed on the outer surface of the phage. This technology is very promising, since one can design an enormous diversity of affinity molecules on the phage's outer surface [72, 73].

3.5 Example applications of electrochemical sensing in the literature

Different viruses have been used as targets in the development of biosensors. At the same time, researchers have focused on detecting either whole viruses, nucleic acids from viruses, or capsid proteins. A large number of publications have described the use of electrochemical biosensors for detecting the presence of pathogenic viruses. Different analytical tools have been used in these studies. Table 3.1 provides an overview of the electrochemical detection technologies used for viruses.

Table 3.1. Overview of electrochemical detection technologies used for viruses.

Virus	Analytical target	Sensor concept/measurement	Detection ability/linear range	References
Avian influenza virus H5N1	DNA and RNA	Genosensor based on SPGEs/ voltammetry	SPGEs: 1 pM, GDEs: 10 pM of RNA transcripts/–	[15]
Avian influenza virus H7	Avian influenza virus H7	Immunosensor based on AgNPs-G/ voltammetry	LOD: 1.6 pg mL^{-1}/1.6 × 10^{-3}~16 ng mL^{-1}	[74]
Dengue	Antibodies	Graphene–polymer/EIS	LOD: 0.12 pfu mL^{-1}/1.0–2.0 × 10^3 pfu mL^{-1}	[75]
Dengue serotype 2	Viral particles	Nanoporous alumina biosensor/EIS	1.0–9.0× 10^2 pfu mL	[76]
Dengue virus	Dengue virus type 2 NS1 protein biomarker	Phage display technology/EIS, CV, SWV	LOD: 0.025 μg mL^{-1}/0.025–3.5 μg mL^{-1}	[77]
Dengue virus	Specific oligonucleotide	DNA based on a nanoporous alumina membrane/impedance	LOD: 2.7 × 10^{-12} M/1.0 × 10^{-12}–1.0 × 10^{-6} M	[78]
Hepatitis B	DNA oligomer	Au-SWCNTs /EIS	LOD: 0.1 pM for one-base mismatched HBV ssDNA	[79]
Human papillomavirus	DNA oligomer	Au-SWCNTs /EIS	LOD: 0.1 pM for one-base mismatched HPV ssDNA/10^{-13}–10^{-8} M	[79]
Human papilloma virus	Human papilloma virus DNA	Graphene/gold nanorod/polythionine/ EIS, DPV	LOD: 4.03 10^{-14} mol L^{-1}/1.0 × $10^{-13-1.0}$ ×10^{-10} mol L^{-1}	[40]
Hepatitis B	Virus surface antigen	Immunosensor based on an Au sheet as working electrode, Fe_3O_4 magnetic nanoparticles/EIS	LOD: 0.19 pg mL^{-1}/0.3–1.0 × 10^3 pg mL^{-1}	[80]
Hepatitis B	Hepatitis B DNA	Nanomaterial-based sensing using surface-immobilized aptamers/EIS, CV, SWV	LOD: 1100 copies/mL/1.10 × 10^3–1.21 × 10^5 copies	[17]
Hepatitis B	Hepatitis B DNA	Streptavidin functionalized magnetic nanoparticles immobilized on a gold electrode/EIS	LOD: 50 pmol/2.53–50.6 nmol mL^{-1}	[81]

(*Continued*)

Table 3.1. (*Continued*)

Virus	Analytical target	Sensor concept/measurement	Detection ability/linear range	References
Hepatitis B	Hepatitis B DNA	Paper-based/potentiometric	LOD: 85 pM/0–1.5 nM	[82]
Hepatitis B	Hepatitis B DNA	Nanoporous gold platform/EIS	LOD: 1×10^{-11} mol/3×10^{-5}–1×10^{-3} M	[18]
Hepatitis B	Hepatitis B antigen	Gold-binding polypeptide fused antibody/EIS	LOD: 0.14 ng ml^{-1}/1 ng/ml -50 µg mL^{-1}	[67]
Influenza virus	Biomarker (M1 protein)	Polyclonal anti-M1 antibodies modified boron-doped diamond/ EIS	LOD: 1 fg mL^{-1} in saliva/up to 100 fg	[83]
Influenza A virus H5N1	Anti-hemagglutinin antibody	Immunosensor based on GCEs	LOD: 2.1 pg mL^{-1}, LOQ: 6.3 pg mL^{-1} 4.0–20.0 pg mL^{-1}	[84]
Influenza virus (type A)	Influenza virus (type A) DNA	DNA immobilized carbon multiwalled nanotubes (MWCNTs)	LOD: 0.5 nM/1–10 nM	[85]
Influenza virus H9N2	Influenza virus H9N2	Receptor-attached AuNPs and MNPs	LOD: 8 HAU/8–128 HAU	[52]
White spot syndrome virus	White spot binding protein	Gold electrode modified with glutathione-S-transferase/ capacitance	–/1 and 1×10^5 copies µL^{-1}	[86]
HIV-1	Capsid protein (p24 antigen)	Immunosensor based on AuNPs/ capacitance	LOD: 7.9×10^{-8} pg mL^{-1}/2.4×10^{-6} to 2.4×10^{-3} pg mL^{-1}	[21]
HIV-1	HIV-1 gene	MIP-based EusNCs/ECL	LOD: 0.3 fm/3.0 fM–0.3 nM	[41]
HIV-1	Capsid protein (p24 antigen)	MIP-based MWCNT-modified GCEs/CV, DPV	LOD: 0.083pgcm^{-3}/1.0×10^{-4}–2ngcm^{-3}	[31]
Human rhinovirus serotype 2	Human rhinovirus serotype 2	MIP-based contactless dielectric microsensors/EIS	–/4 µg mL^{-1}–3 µg mL^{-1}	[87]
Zika virus	Virus protein	Immunosensor based on interdigitated gold microelectrode array/EIS	LOD: 10 pm/10 pm to 1 nM	[47]

Zika virus	Zika virus	Surface imprinting-based graphene oxide composite/EIS	LOD: 2×10^{-4} PFU mL^{-1}/10^{-3-102} PFU mL^{-1}	[32]
Zika virus	Zika virus NS1 antigen	Mabs immobilized graphene	LOD: 0.45 nM/500 ng mL^{-1}	[88]
Ebola	Ebola DNA	Thiolated DNA probe immobilized SPGEs/EIS	LOD: 4.7 nM/10–75 nM	[45]
Bacteriophage	Bacteriophage	MIP-based gold electrode/capacitance	LOD: 10 PFU mL^{-1}/10^{1}–105 PFU mL^{-1}	[89]
Norovirus	Norovirus DNA	DNA conjugated Au/MNP-CNTs	LOD: 8.8 pM/1 pM–10 nM	[90]
MERS-CoV	Recombinant spike protein S1	Immunosensor based on AuNPs on carbon/SWV	LOD: 1.0 pg mL^{-1}/0.001–100 ng ml^{-1}	[91]
SARS-CoV-2	Spike protein	SpAb/AuNPs/FDTO/DPV	LOD: 90 fM	[92]

SPGEs: screen-printed gold electrodes, GDEs: gold disc electrodes, AgNPs-G: silver nanoparticles–graphene, LOD: limit of detection, EIS: electrochemical impedance spectroscopy, Au-SWCNTs: gold nanoparticle–single walled carbon nanotube, LOQ: limit of quantification, GCEs: glassy carbon electrodes, AuNPs: gold nanoparticles, EsNCs: europium sulfide nanocrystals, ECL: electrochemiluminescence, CV: cyclic voltammetry, SWV: square wave voltammetry, MWCNTs: multiwalled carbon nanotubes, DPV: differential pulse voltammetry, PFU: plaque-forming unit, Mabs: monoclonal antibodies, Au/MNP-CNTs: gold/iron-oxide magnetic nanoparticle-decorated carbon nanotubes, MNPs: magnetic nanoparticles, HAU: hemagglutinin

3.6 Future perspectives and conclusions

Due to the heterogeneous nature of viral infections, their early and accurate diagnosis has attracted great interest. Detection platforms with the ability to simultaneously consider multiple factors are key tools. The limitations of virus sensing can be listed as follows: the presence of low concentrations of viruses in complex biological samples, obtaining low LOD values with a high regression coefficient, creating inter- and intra-laboratory applicable sensing platforms. From this point of view, automated biosensors with high analytical detection performance that attract all the attention of the market continue to receive scientific and commercial interest daily. In addition, completely electronic barcoding allows researchers to fit biosensors onto microchips. Lab-on-a-chip has been identified as a very innovative technology, in which the functions performed by large and bulky devices can be performed by a small chip. The developed devices are palm-sized, portable, and can be connected to smartphones. Analysis can be performed in seconds using saliva samples, the results are evaluated using the developed software, and the doctor is immediately notified. These devices can bring significant changes to the medical world as a result of of their features, such as their low cost and ease of use compared to existing methods. Dangerous infections such as HIV, Lyme disease, malaria, and coronavirus could be easily detected by these promising tools. However, some difficulties have continued to be reported; overcoming these by research into the design of new portable biosensors has become crucial. Rapid progress is being achieved in this field, but more investigations of mobile diagnostics based on smartphone technology are needed. It is estimated that these platforms will create a new revolution in global health protection as a result of the widespread use of these tools.

References

[1] Castillo-Henriquez L *et al* 2020 Clinical pathogens *Sensors* **20** 1–26

[2] Ozer T, Geiss B J and Henry C S 2020 Review – chemical and biological sensors for viral detection *J. Electrochem. Soc.* **167** 037523

[3] Chowdhury A D, Takemura K, Li T C, Suzuki T and Park E Y 2019 Electrical pulse-induced electrochemical biosensor for hepatitis E virus detection *Nat. Commun.* **10** 4–7

[4] Imran S, Ahmadi S and Kerman K 2021 Electrochemical biosensors for the detection of SARS-CoV-2 and other viruses *Micromachines* **12** 1–23

[5] Chan Y K and Gack M U 2016 Viral evasion of intracellular DNA and RNA sensing *Nat. Rev. Microbiol.* **14** 360–73

[6] Bergdahl G E, Hedström M and Mattiasson B 2019 Capacitive sensor to monitor enzyme activity by following degradation of macromolecules in real time *Appl. Biochem. Biotechnol.* **189** 374–83

[7] Saylan Y, Erdem Ö, Ünal S and Denizli A 2019 An alternative medical diagnosis method: biosensors for virus detection *Biosensors* **9** 65

[8] Ertürk G and Mattiasson B 2017 Molecular imprinting techniques used for the preparation of biosensors *Sensors* **17** 288

[9] Idil N and Mattiasson B 2017 Imprinting of microorganisms for biosensor applications *Sensors* **17** 1–15

[10] Ahmed A, Rushworth J V, Hirst N A and Millner P A 2014 Biosensors for whole-cell bacterial detection *Clin. Microbiol. Rev.* **27** 631–46

[11] Nguyen H H, Yi S Y, Woubit A and Kim M 2016 A portable surface plasmon resonance biosensor for rapid detection of *Salmonella typhimurium Appl. Sci. Converg. Technol.* **25** 61–5

[12] Idil N, Hedström M, Denizli A and Mattiasson B 2016 Whole cell based microcontact imprinted capacitive biosensor for the detection of *Escherichia coli Biosens. Bioelectron.* **87** 807–15

[13] Erdem Ö, Saylan Y, Cihangir N and Denizli A 2019 Molecularly imprinted nanoparticles based plasmonic sensors for real-time *Enterococcus faecalis* detection *Biosens. Bioelectron.* **126** 608–14

[14] Dar K K, Shao S, Tan T and Lv Y 2020 Molecularly imprinted polymers for the selective recognition of microorganisms *Biotechnol. Adv.* **45** 107640

[15] Malecka K *et al* 2016 Electrochemical genosensor based on disc and screen printed gold electrodes for detection of specific DNA and RNA sequences derived from Avian Influenza Virus H5N1 *Sensors Actuators* B **224** 290–7

[16] Samson R, Navale G R and Dharne M S 2020 Biosensors: frontiers in rapid detection of COVID-19 *3 Biotech.* **10** 1–9

[17] Zhao F *et al* 2020 New electrochemical DNA sensor based on nano flowers of $Cu_3(PO_4)_2$-BSA-GO for hepatitis B virus DNA detection *J. Electroanal. Chem.* **867** 114184

[18] Ahangar L E and Mehrgardi M A 2017 Amplified detection of hepatitis B virus using an electrochemical DNA biosensor on a nanoporous gold platform *Bioelectrochemistry* **117** 83–8

[19] Rodríguez Y, Novelli L, Rojas M, De Santis M, Acosta-Ampudia Y, Monsalve D M and Ramírez-Santana C *et al* 2020 Autoinflammatory and autoimmune conditions at the crossroad of COVID-19 *J. Autoimmun.* **114** 102506

[20] Singh P 2017 Surface plasmon resonance: a boon for viral *Reference Module in Life Sciences* (https://doi.org/10.1016/B978-0-12-809633-8.12245-9)

[21] Teeparuksapun K, Hedstro M, Wong E Y, Tang S, Hewlett I K and Mattiasson B 2010 Letters to analytical chemistry ultrasensitive detection of HIV-1 p24 antigen using nano-functionalized surfaces in a capacitive immunosensor *Anal. Chem.* **82** 8406–11

[22] Nourani S, Ghourchian H and Boutorabi S M 2013 Magnetic nanoparticle-based immunosensor for electrochemical detection of hepatitis B surface antigen *Anal. Biochem.* **441** 1–7

[23] Wang J 2002 Electrochemical nucleic acid biosensors *Anal. Chim. Acta* **469** 63–71

[24] Ju H X, Ye Y K, Zhao J H and Zhu Y L 2003 Hybridization biosensor using di(2, 2′-bipyridine)osmium (III) as electrochemical indicator for detection of polymerase chain reaction product of hepatitis B virus DNA *Anal. Biochem.* **313** 255–61

[25] Khan M Z H, Hasan M R, Hossain S I, Ahommed M S and Daizy M 2020 Ultrasensitive detection of pathogenic viruses with electrochemical biosensor: state of the art *Biosens. Bioelectron.* **166** 112431

[26] Tepeli Y and Ülkü A 2018 Electrochemical biosensors for influenza virus a detection: the potential of adaptation of these devices to POC systems *Sensors Actuators* B **254** 377–84

[27] Ertürk G, Berillo D, Hedström M and Mattiasson B 2014 Microcontact-BSA imprinted capacitive biosensor for real-time, sensitive and selective detection of BSA *Biotechnol. Reports* **3** 65–72

[28] Rebelo P *et al* 2021 Molecularly imprinted polymer-based electrochemical sensors for environmental analysis *Biosens. Bioelectron.* **172** 112719

[29] Cui F, Zhou Z and Zhou H S 2020 Molecularly imprinted polymers and surface imprinted polymers based electrochemical biosensor for infectious diseases *Sensors* **20** 996

[30] Chow E J and Mermel L A 2017 Hospital-acquired respiratory viral infections: incidence, morbidity, and mortality in pediatric and adult patients *Open Forum Infect. Dis.* **4** 1–6

[31] Ma Y, Shen X L, Zeng Q, Wang H S and Wang L S 2017 A multi-walled carbon nanotubes based molecularly imprinted polymers electrochemical sensor for the sensitive determination of HIV-p24 *Talanta* **164** 121–7

[32] Tancharoen C, Sukjee W, Thepparit C, Jaimipuk T, Auewarakul P and Thitithanyanont A *et al* 2019 Electrochemical biosensor based on surface imprinting for Zika virus detection in serum *ACS Sens.* **4** 69–75

[33] Agrawal R 2021 Hepatitis C nobel prize: the exhumation *Indian J. Pathol. Microbiol.* **64** 1

[34] Esparza-Leal H M *et al* 2009 Detection of white spot syndrome virus in filtered shrimp-farm water fractions and experimental evaluation of its infectivity in Penaeus (Litopenaeus) vannamei *Aquaculture* **92** 16–22

[35] Burrell C J, Howard C R and Murphy F A 2017 Coronaviruses *Fenner and White's Medical Virology* 5th edn (New York: Academic) pp 437–69

[36] Pal M, Berhanu G, Desalegn C and Kandi V 2020 Severe acute respiratory syndrome Coronavirus-2 (SARS-CoV-2): an update *Cureus* **12** e7423

[37] Greenhalgh T, Knight M, Inada-kim M, Fulop N J, Leach J and Vindrola-padros C 2021 Remote management of Covid-19 using home pulse oximetry and virtual ward support *Brit. Med. J.* **372**

[38] Yakoh A, Pimpitak U, Rengpipat S and Hirankarn N 2021 Paper-based electrochemical biosensor for diagnosing COVID-19: detection of SARS-CoV-2 antibodies and antigen *Biosen. Bioelectron.* **176** 112912

[39] Chaibun T, Puenpa J, Ngamdee T, Boonapatcharoen N, Athamanolap P and Mullane A P O *et al* 2019 Rapid electrochemical detection of coronavirus SARS-CoV-2 *Nat. Commun.* **2021** 1–10

[40] Alafeef M, Dighe K, Moitra P and Pan D 2020 Rapid, ultrasensitive, and quantitative detection of SARS-CoV-2 using antisense oligonucleotides directed electrochemical biosensor chip *ACS Nano* **14** 17028–45

[41] Babamiri B, Salimi A and Hallaj R 2018 A molecularly imprinted electrochemiluminescence sensor for ultrasensitive HIV-1 gene detection using EuS nanocrystals as luminophore *Biosens. Bioelectron.* **117** 332–39

[42] Gong Q *et al* 2019 Sensitive electrochemical DNA sensor for the detection of HIV based on a polyaniline/graphene nanocomposite *J. Mater.* **5** 313–9

[43] Balbin Tamayo A I, Lopez Rizo L S, de Armas M B, Yamanaka H and Esteva Guas A M 2018 Free-label electrochemical detection of of HIV-1 proviral DNA in clinical sample *Int. J. Biosens. Bioelectron.* **5** 217–9

[44] Yeter E, Şahin S, Caglayan M O and Üstündağ Z 2021 An electrochemical label-free DNA impedimetric sensor with AuNP-modified glass fiber/carbonaceous electrode for the detection of HIV-1 DNA *Chem. Pap.* **75** 77–87

[45] Ilkhani H and Farhad S 2018 A novel electrochemical DNA biosensor for Ebola virus detection *Anal. Biochem.* **557** 151–5

[46] Hennessey M, Fischer M and Staples J E 2016 Zika virus spreads to new areas – region of the Americas *MMWR Morb. Mortal Wkly. Rep.* **65** 1–4
[47] Kaushik A *et al* 2018 A sensitive electrochemical immunosensor for label-free detection of Zika-virus protein *Sci. Rep.* **8** 9700
[48] Borkenhagen L K, Salman M D, Ma M J and Gray G C 2019 Animal influenza virus infections in humans: a commentary *Int. J. Infect. Dis.* **88** 113–9
[49] Lakshmipriya T, Gopinath S C B and Hashim U 2018 Influenza viral detection on microfluidic delivery assisted biosensors *Microsyst. Technol.* **24** 823–30
[50] Richmond H, Rees N, McHale S, Rak A and Anderson J 2020 Seasonal influenza vaccination during a pandemic *Hum. Vaccines Immunother.* **16** 2219–21
[51] Diouani M F *et al* 2008 Miniaturized biosensor for avian influenza virus detection *Mater. Sci. Eng.* C **28** 580–3
[52] Sayhi M *et al* 2018 Electrochemical detection of influenza virus H9N2 based on both immunomagnetic extraction and gold catalysis using an immobilization-free screen printed carbon microelectrode *Biosens. Bioelectron.* **107** 170–7
[53] Lakard S, Pavel I and Lakard B 2021 Electrochemical biosensing of dopamine neurotransmitter: a review *Biosensors* **11** 179
[54] Amiri M, Bezaatpour A, Jafari H, Boukherroub R and Szunerits S 2018 Electrochemical methodologies for the detection of pathogens *ACS Sens.* **3** 1069–86
[55] Lopes F, Pacheco J G, Rebelo P and Delerue-Matos C 2017 Molecularly imprinted electrochemical sensor prepared on a screen printed carbon electrode for naloxone detection *Sensors Actuators* B **243** 745–52
[56] Raffaelle R P, Landi B J, Harris J D, Bailey S G and Hepp A F 2005 Carbon nanotubes for power applications *Mater. Sci. Eng.* B **116** 233–43
[57] Huang H, Bai W, Dong C, Guo R and Liu Z 2015 An ultrasensitive electrochemical DNA biosensor based on graphene/Au nanorod/polythionine for human papillomavirus DNA detection *Biosens. Bioelectron.* **68** 442–6
[58] Han J, Lee D, Hong C, Chew C, Kim T and Jungho J 2016 A multi-virus detectable microfluidic electrochemical immunosensor for simultaneous detection of H1N1, H5N1, and H7N9 virus using ZnO nanorods for sensitivity enhancement *Sensors Actuators* B **228** 36–42
[59] Cox K L *et al* 2012 Immunoassay Methods: Immunoassay development, optimization and validation flow chart (https://www.ncbi.nlm.nih.gov/books/NBK92434/)
[60] Sakamoto S *et al* 2018 Enzyme-linked immunosorbent assay for the quantitative/qualitative analysis of plant secondary metabolites *J. Nat. Med.* **72** 32–42
[61] Shan G, Huang W, Gee S J, Buchholz B A, Vogel J S and Hammock B D 2000 Isotope-labeled immunoassays without radiation waste *Proc. Natl. Acad. Sci. USA* **97** 2445–9
[62] Cho I H, Kim D H and Park S 2020 Electrochemical biosensors: perspective on functional nanomaterials for on-site analysis *Biomater. Res.* **24** 1–12
[63] Prodromidis M I 2010 Impedimetric immunosensors: a review *Electrochim. Acta* **55** 4227–33
[64] Waiyapoka T *et al* 2015 Application of a label-free immunosensor for White Spot Syndrome Virus (WSSV) in shrimp cultivation water *Appl. Biochem. Biotechnol.* **177** 821–30
[65] Bart M, Stigter E C A, Stapert H R, De Jong G J, Van and Bennekom W P 2005 On the response of a label-free interferon-γ immunosensor utilizing electrochemical impedance spectroscopy *Biosens. Bioelectron.* **21** 49–59
[66] Bogomolova A *et al* 2009 Challenges of electrochemical impedance spectroscopy in protein biosensing *Anal. Chem.* **81** 3944–9

[67] Heo N S *et al* 2012 Label-free electrochemical diagnosis of viral antigens with genetically engineered fusion protein *Sensors* **12** 10097–108
[68] Hoon S *et al* 2019 Development of a rapid and sensitive electrochemical biosensor for detection of human norovirus via novel speci fi c binding peptides *Biosens. Bioelectron.* **123** 223–9
[69] Erofeev A S, Gorelkin P V, Kolesov D V, Kiselev G A, Dubrovin E V and Yaminsky I V 2019 Label-free sensitive detection of influenza virus using PZT discs with a synthetic sialylglycopolymer receptor layer *R. Soc. Open Sci.* **6** 1–13
[70] Zuo B, Li S, Guo Z, Zhang J and Chen C 2004 Piezoelectric immunosensor for SARS-associated coronavirus in sputum *Anal. Chem.* **76** 3536–40
[71] Bisoffi M, Severns V, Branch D W, Edwards T L and Larson R S 2013 Rapid detection of human immunodeficiency virus types 1 and 2 by use of an improved piezoelectric biosensor *J. Clin. Microbiol.* **51** 1685–91
[72] Jung S, Honegger A and Plückthun A 1999 Selection for improved protein stability by phage display *J. Mol. Biol.* **294** 163–80
[73] Azzazy H M E and Highsmith W E 2002 Phage display technology: clinical applications and recent innovations *Clin. Biochem.* **35** 425–45
[74] Huang J *et al* 2016 Silver nanoparticles coated graphene electrochemical sensor for the ultrasensitive analysis of avian influenza virus H7 *Anal. Chim. Acta* **913** 121-27
[75] Navakul K, Warakulwit C, Yenchitsomanus P-thai, Panya A, Lieberzeit P A and Sangma C 2017 A novel method for dengue virus detection and antibody screening using a graphene-polymer based electrochemical biosensor *Nanomedicine: NBM* **13** 549–57
[76] Thi B *et al* 2012 Bioelectrochemistry Electrochemical impedance spectroscopy characterization of nanoporous alumina dengue virus biosensor *Bioelectrochemistry* **88** 15–21
[77] Lim J M, Kim J H, Ryu M Y, Cho C H, Park T J and Park J P 2018 An electrochemical peptide sensor for detection of dengue fever biomarker *Anal. Chim. Acta* **1026** 109–16
[78] Deng J and Toh C S 2013 Impedimetric DNA biosensor based on a nanoporous alumina membrane for the detection of the specific oligonucleotide sequence of dengue virus *Sensors* **13** 7774–85
[79] Wang S *et al* 2013 Electrochemical detection of hepatitis B and papilloma virus DNAs using SWCNT array coated with gold nanoparticles *Biosens. Bioelectron.* **41** 205–10
[80] Alizadeh N, Hallaj R and Salimi A 2017 A highly sensitive electrochemical immunosensor for hepatitis B virus surface antigen detection based on Hemin/G-quadruplex horseradish peroxidase-mimicking DNAzyme-signal amplification *Biosens. Bioelectron.* **94** 184–92
[81] Hassen W M, Chaix C, Abdelghani A, Bessueille F, Leonard D and Jaffrezic-Renault N 2008 An impedimetric DNA sensor based on functionalized magnetic nanoparticles for HIV and HBV detection *Sensors Actuators* B **134** 755–60
[82] Li X, Scida K and Crooks R M 2015 Detection of Hepatitis B virus DNA with a paper electrochemical sensor *Anal. Chem.* **87** 9009–15
[83] Nidzworski D *et al* 2017 A rapid-response ultrasensitive biosensor for influenza virus detection using antibody modified boron-doped diamond *Sci. Rep.* **7** 1–10
[84] Jarocka U *et al* 2014 Electrochemical immunosensor for detection of antibodies against influenza A virus H5N1 in hen serum *Biosens. Bioelectron.* **55** 301–6
[85] Tam P D, Van Hieu N, Chien N D, Le A T, Anh and Tuan M 2009 DNA sensor development based on multi-wall carbon nanotubes for label-free influenza virus (type A) detection *J. Immunol. Methods* **350** 118–24

[86] Samanman S, Kanatharana P, Chotigeat W and Deachamag P 2011 Highly sensitive capacitive biosensor for detecting white spot syndrome virus in shrimp pond water *J. Virol. Methods* **173** 75–84

[87] Birnbaumer G M *et al* 2009 Detection of viruses with molecularly imprinted polymers integrated on a microfluidic biochip using contact-less dielectric microsensors *Lab Chip* **9** 3549–56

[88] Afsahi S *et al* 2018 Novel graphene-based biosensor for early detection of Zika virus infection *Biosens. Bioelectron.* **100** 85–8

[89] Ertürk G and Lood R 2018 Bacteriophages as biorecognition elements in capacitive biosensors: phage and host bacteria detection *Sensors Actuators* B **258** 535–43

[90] Lee J, Morita M, Takemura K and Park E Y 2018 A multi-functional gold/iron-oxide nanoparticle-CNT hybrid nanomaterial as virus DNA sensing platform *Biosens. Bioelectron.* **102** 425–31

[91] Layqah L A and Eissa S 2019 An electrochemical immunosensor for the corona virus associated with the Middle East respiratory syndrome using an array of gold nanoparticle-modified carbon electrodes *Microchim. Acta* **186** 224

[92] Mahari S, Roberts A, Shahdeo D and Gandhi S 2020 Ecovsens-ultrasensitive novel in-house built printed circuit board based electrochemical device for rapid detection of nCovid-19 antigen, a spike protein domain 1 of SARS-CoV-2 *bioRxiv* (https://doi.org/10.1101/2020.04.24.059204v3)

IOP Publishing

Biosensors for Virus Detection

Adil Denizli and Yeşeren Saylan

Chapter 4

Piezoelectric biosensors for virus detection

Semra Akgönüllü and Adil Denizli

Viruses are pathogenic microorganisms that cause many infectious diseases. Viruses can inhabit the human body and replicate in it; moreover, they can spread easily and quickly from infected people to healthy people. For this reason, the most important way to address the undesirable spread and subsequent medical treatment of infectious illnesses is to detect the viruses that cause such diseases in a timely manner. Biosensors are defined as smart analytical tools that incorporate biological recognition elements or biomimetics, which are used for the quick detection and real-time monitoring of molecules that are significant in the fields of medicine, security, food, and the environment. Piezoelectric biosensors are mass-sensitivity-based micromechanical transducers. Their mass is increased by interactions between molecules and the frequency is checked using alternate current and voltage decreases. This chapter summarizes the importance of piezoelectric biosensors and offers an in-depth discussion of the use of piezoelectric biosensors in the detection of viruses.

4.1 Introduction

Typical chemical sensors have been utilized to find, detect, or quantify substances and energy, or supply a signal to confirm the detection of chemical or physical features of interest [1, 2]. A biosensor is generally the same as a sensor, except that the recognition material is a biological component. Biosensors are intended to detect biochemical or biological reactions. The sensing surface of a biosensor can be designed to take various forms [3–5]. Biosensors offer sensitivity, fast response time, robustness, and portability. They are a potential alternative to traditional grade-evaluation devices, but with easier protocols. Biosensors must address the requirements for reproducibility, sensitivity, response correctness, nontoxicity, high specificity toward the target template molecule, and cost-effectiveness [6]. Biosensors offer rapid diagnosis and better health care. Biosensors can help to reduce the waiting time for test results, which is highly stressful for patients [7]. Biosensors can be separated into various transducer groups based on the method of generation of

doi:10.1088/978-0-7503-3867-7ch4 © IOP Publishing Ltd 2021

the output signal, which can be electrochemical, optical, magnetic, gravimetric, or piezoelectric [6, 8–11]. The selection of a transduction tool is primarily based on the natural and physicochemical features of its surface layer, which change as a result of interaction with the target molecule [12].

Figure 4.1 shows a general view of the basic approaches to the use of transducers in biosensors. Optical transducers can determine the presence of molecules by measuring changes in absorption, plasmons, or fluorescence [13]. Ren *et al* developed a fluorescent biosensor for sensitive analysis of the acetylcholinesterase (AChE) reaction system based on carbon dots. Figure 4.1(a) shows a schematic of a fluorescent biosensor system designed for the detection of H_2O_2-related substrates and enzymes based on carbon dots [14]. Electrochemical biosensors can be classified as impedimetric, amperometric, conductometric, or potentiometric. Figure 4.1(b) shows a simple label-free electrochemical DNA biosensor [15]. Piezoelectric transducers measure change in the frequency of resonance. A piezoelectric-based quartz crystal microbalance (QCM) utilizes the frequency change of the sensing electrode (figure 4.1(c)) [16, 17]. There are other biosensor systems, apart from the three biosensors shown. Other tools, such as field-effect transistor-based, console-based, and microfluidic biosensors have also been fabricated and studied in various fields.

Electromechanical transducers based on piezoelectric materials have been installed as biosensors and actuators in vehicles and smart structures. Piezoelectric, mass-sensitive, or acoustic tools enable label-free and quick detection of molecules based on the molecular mass that is a characteristic feature of a molecule. These transducers include: (a) surface acoustic wave (SAW) transducers [18, 19]; (b) bulk acoustic wave (BAW) devices, such as QCMs [20–22]; and (c) shear transverse wave (STW) devices [23], which are generally used in combination with various bioreceptor elements for biomedical diagnosis and molecular recognition [24].

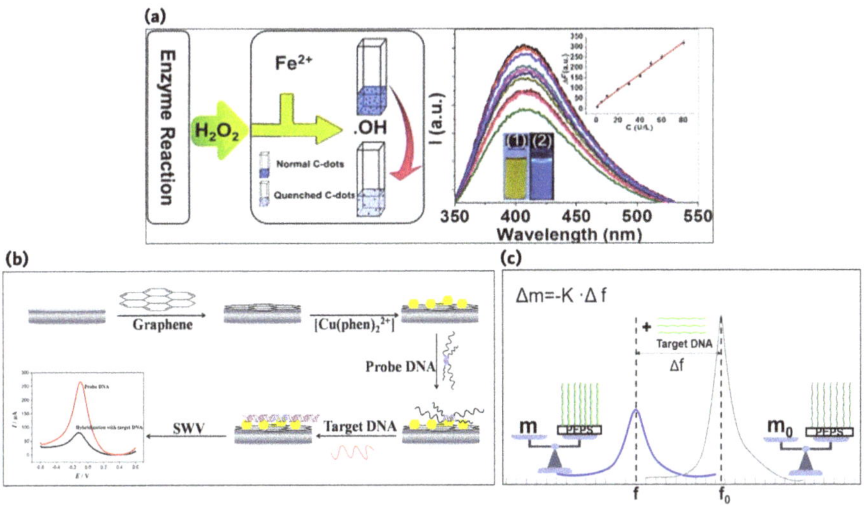

Figure 4.1. Biosensor systems: (a) an optical biosensor [14]; (b) an electrochemical biosensor [15]; (c) a piezoelectric biosensor [17]. (a) Reprinted from [14], Copyright 2015, with permission from Elsevier. (b) Reprinted from [15], Copyright 2016, with permission from Elsevier. (c) Reprinted from [17], Copyright 2020, with permission from Elsevier.

Viruses are pathogenic microorganisms that cause many infectious diseases, such as hepatitis, gastroenteritis, meningitis, influenza, acquired immune deficiency syndrome (AIDS), pneumonia, etc [24]. Human immunodeficiency virus (HIV), avian influenza, Ebola virus, and severe acute respiratory syndrome coronavirus (SARS) are viruses that originate from wild animals and are examples of zoonoses that pose a increasingly critical threat to human health and global economies [4, 16]. Detecting the viruses that cause such diseases in time is the most significant step in the treatment and spread of these infectious diseases. Piezoelectric transducers based on natural or synthetic receptors for viral diagnostics are used as portable biosensors that can sense and selectively identify and quantify dangerous virus genes. This chapter describes the design of a novel piezoelectric biosensor for the detection of 100 nm viruses, including the hepatitis B virus (HBV), the hepatitis C virus (HCV), the dengue virus, influenza, Ebola virus, vaccinia virus, human papillomavirus (HPV), human immunodeficiency virus (HIV), and SARS-associated coronavirus (SARS-CoV). This chapter also summarizes the viruses commonly targeted by piezoelectric biosensors and emphasizes the main relevant clinical parameters, such as the detection limit, concentration range, and bioassay design.

4.2 Piezoelectric-based biosensors

There is a linear relationship between a change in the oscillation frequency of a piezoelectric crystal and a mass-sensitive change in metal-based electrodes. The most well-known oscillatory biosensors are the BAW and SAW tools. BAW tools work according to the Sauerbrey equation shown below [26]. The Sauerbrey equation describes the relationship between the resonance frequency of a crystal and the mass deposited on its electrodes. The use of piezoelectric devices as potential biosensors is based on the Sauerbrey equation:

$$\Delta F = -2.3 \times 10^6 F^2 \Delta M / A$$

where F is the resonant frequency of the crystal, ΔF is the change in frequency, ΔM is the mass deposited, and A is the field coated.

Sauerbrey's equation is also utilized to correlate the mass variation at the QCM crystal chip's surface with the change in the oscillation frequency of the QCM crystal. As a result of the interactions between molecules, the mass m increases, and as a result of this, the frequency f controlled by the alternating current voltage decreases:

$$\Delta F = -C_f \times \Delta_m$$

King first described the analytical technique of the adsorption-sensing crystal detector [27]. The first applications were measurements in the gas phase. Technological advances that later facilitated measurements of the liquid phase release created the rapidly developing field of piezoelectric-based biosensors [28]. Various works have been published concerning the theoretical directions of oscillation in solution, including research addressing solution viscosity and the frequency shift due to density [29].

Schematic diagrams of the procedural basis of the piezoelectric-based biosensor system and a graph of the time-dependent change of frequency are presented in figure 4.2. Specific antibodies are attached to the upper metal surface (e.g. gold), as displayed in figure 4.2(a). The lower and upper piezoelectric electrodes resonate the piezoelectric biosensor. The target antigens then bind specifically to antibodies. The effect of the change of mass Δm of the electrode's surface on the change in the frequency f of the material in the oscillating circuit is given in the time-dependent graph in figure 4.2(c).

Piezoelectric materials include ceramics, crystals, polymers such as polyvinylidene fluoride, and piezoelectric composites. Piezoelectric materials convert mechanical energy into electrical energy and vice versa [30]. Generally, anisotropic materials such as polyvinylidene fluoride (PVDF) [31, 32], zinc oxide (ZnO) [33], aluminum nitride (AlN) [34, 35], barium titanate ($BaTiO_3$) [36, 37], quartz (SiO_2) [38, 39], and lead titanate ($PbTiO_3$) [40, 41] are utilized as biosensor materials [42]. Piezoelectric biosensor devices based on label-free techniques have made great advances [43] and have been successfully used for virus detection. A schematic diagram of a quartz crystal and photographs of three different quartz crystals (AT-cut 10 MHz) are shown in figure 4.3 [44, 45].

Thus, piezoelectric-based biosensors are very popular in the areas of food control [46], drug delivery mechanisms [47, 48], environmental pollution [49], etc. Piezoelectric biosensors, which are mass-based chemical sensors (e.g. QCMs), have important properties, such as high sensitivity and selectivity, ease of use, cost-effectiveness, stability, portability, and simplicity. Several methods can be used to design piezoelectric biosensor surfaces for different application fields [48]. The widespread use of piezoelectric-based biosensors has been documented for the detection of various analytes such as proteins [50], enzymes [51], drugs [52], pesticides [53], metals [54], antibiotics [55], biomarkers [56], bacteria [57], and vitamins [58], etc. The mass-sensitive type of piezoelectric-based biosensor has commonly been utilized for the detection of viruses.

4.3 Virus detection applications

Mass-sensitive piezoelectric devices comprise a significant group of biosensors. Shons made the first piezoelectric biosensor to detect immunoglobulin G antibodies from cow serum in 1972. Since then, biosensors have been significant in biological detection because of their simplicity, specificity, sensitivity, and rapidity [59]. Here,

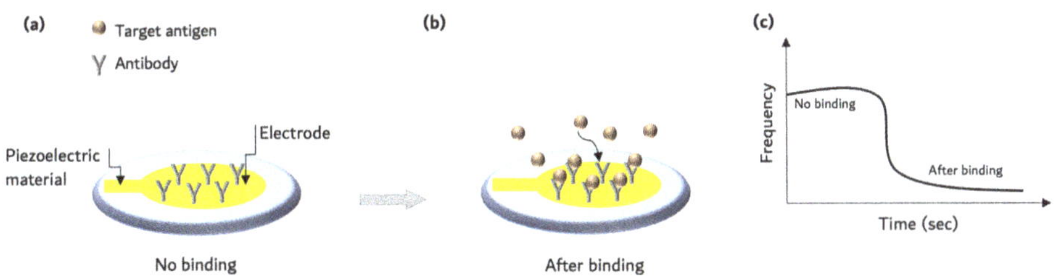

Figure 4.2. (a) The general design of virus detectors based on piezoelectric electrodes; (b) working method of piezoelectric biosensors; and (c) change of frequency due to time-dependent detection.

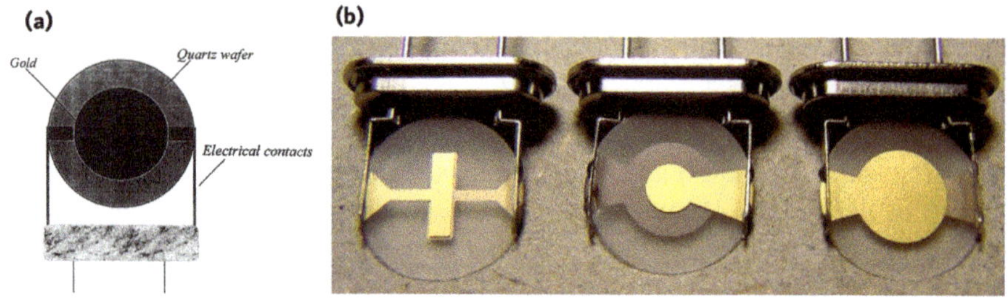

Figure 4.3. (a) Schematic diagram of a typical piezoelectric crystal [44]; (b) AT-cut 10 MHz quartz crystal rectangular and circular electrodes [45]. (a) Reprinted from [44], Copyright 1999, with permission from Elsevier. (b) Reprinted from [45], Copyright 2007, with permission from Elsevier.

we describe state-of-the-art investigations in which piezoelectric-based biosensor devices have been examined for use in virus detection.

4.3.1 Hepatitis B and C viruses

The hepatitis B virus (HBV) causes great problems for human health worldwide. It can cause chronic hepatitis, which easily leads to cirrhosis and primary liver cancer [60]. Xu *et al* fabricated a piezoelectric immunoassay biosensor for the real-time detection of anti-alpha fetoprotein (AFP) and anti-hepatitis B virus (HBV). The piezoelectric sensing material was produced by the micro-machining technique and possessed eight equal circular surfaces. The limit of detection was found to be 0.1 ng mL^{-1} and the linear concentration range was between 0.1 and 10 000 ng mL^{-1} for anti-HBsAg and anti-AFP. This piezoelectric chip has the potential to be applied for quick, selective, and repeated detection of the hepatitis B virus [61].

Zhou *et al* established a highly sensitive piezoelectric biosensor for the detection of HBV DNA. HBV nucleic acid (HBV DNA) was attached to the gold surface of a piezoelectric quartz crystal (9 MHz AT-cut). The concentration range was 0.02–0.14 mg mL^{-1}. The designed crystal was used almost five times without a noticeable decrease in sensitivity. The reusability, selectivity, and sensitivity of this developed QCM biosensor promise a bright future for such sensors in further applications [62].

The hepatitis C virus (HCV) is a potential cause of viral hepatitis, which leads to liver cirrhosis, hepatocellular carcinoma, and blood-borne chronic hepatitis [63]. Skládal *et al* established a piezoelectric quartz crystal chip for the detection of the hepatitis C virus in serum samples. They modified piezoelectric quartz crystal resonators using oligonucleotide probes. Piezoelectric-based biosensors seem to be a fit and useful vehicle for the real-time monitoring of oligonucleotides. Piezoelectric-based biosensors can reduce the assay procedure to 10 min and lessen the analysis time. The samples of DNA were taken from the commercial analysis of the Amplicor hepatitis C virus, which has a total length of 244 bases [64].

4.3.2 Influenza virus

While old viruses are getting stronger day by day, new viruses also cause epidemics. Therefore, current biosensing systems need continuous updates to address the

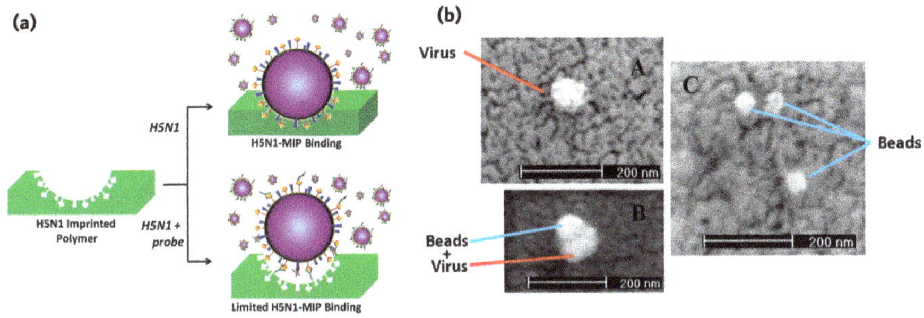

Figure 4.4. (a) Binding of the H5N1 virus to an MIP surface [67] and (b) scanning electron microscope images of the top view of the gold surface of the QCM electrode; (A) H5N1 virus; (B) magnetic nanoparticle and H5N1 virus; (C) magnetic nanoparticles [68]. (a) Reproduced from [67] with permission of The Royal Society of Chemisty. (b) Reprinted from [68], Copyright 2011, with permission from Elsevier.

strongly growing challenges of viral diagnosis. There are four types of influenza virus, types A, B, C, and D. Influenza types A and B are responsible for major outbreaks, in contrast to types C and D [42, 65, 66].

Wangchareansak *et al* designed a molecularly imprinted polymer (MIP)-coated QCM biosensor for the detection of an inactivated strain of influenza A H5N1 [67]. Figure 4.4(a) displays a schematic diagram of the H5N1 virus binding to the MIP surface. Hemagglutinin (HA) and neuraminidase (NA) are the two main proteins found on the H5N1 surface. Methacrylic acid and methyl-methacrylate were used as the functional monomer and the monomer, respectively. H5N1 was used as a template molecule. Sample injection was performed using a buffer solution with a pH of 7.2.

Avian influenza (AI) H5N1 is highly pathogenic. Quick detection of the H5N1 virus is crucial to preventing the spread of a potential pandemic threat to human health. A novel method involving the use of a magnetic nanoparticle amplification-based QCM biosensor for the detection of the AI H5N1 virus was developed by Li *et al.* They immobilized polyclonal antibodies against AI H5N1 virus surface antigen HA on the QCM crystal gold surface; 16-mercaptohexadecanoic acid (MHDA) was used for the self-assembled monolayer process. SEM images of the H5N1 virus, magnetic nanoparticles and an H5N1 virus, and magnetic nanoparticles are shown in figure 4.4(b) [68].

H1N1 is known as swine flu and is a member of the *Orthomyxoviridae* family. H1N1 is caused by the influenza A virus. It is spread as a result of the transmission of H1N1 from person to person in the air, or sometimes from pigs to humans [65, 66]. Jiang *et al* designed and fabricated an SAW biosensor for the sensitive detection of the HA of the influenza A H1N1 virus subtype. Influenza A H1N1 HA antibodies were successfully immobilized. The SAW biosensor electrode surface was characterized using Fourier-transform infrared spectroscopy (FTIR) analysis. SAW biosensors detected HA to a concentration of as little as 1 ng ml^{-1}. An image of the SAW biosensor setup is shown in figures 4.5(a) and (b) [69]. In another study, Erofeev developed a method for the label-free and quick detection of influenza A viruses in air or a water solution. Lead zirconate titanate (PZT) piezoelectric discs were used in the first radial mode. Piezoelectric membrane discs were fabricated with a radius of 2 mm and a thickness of 100μm. They

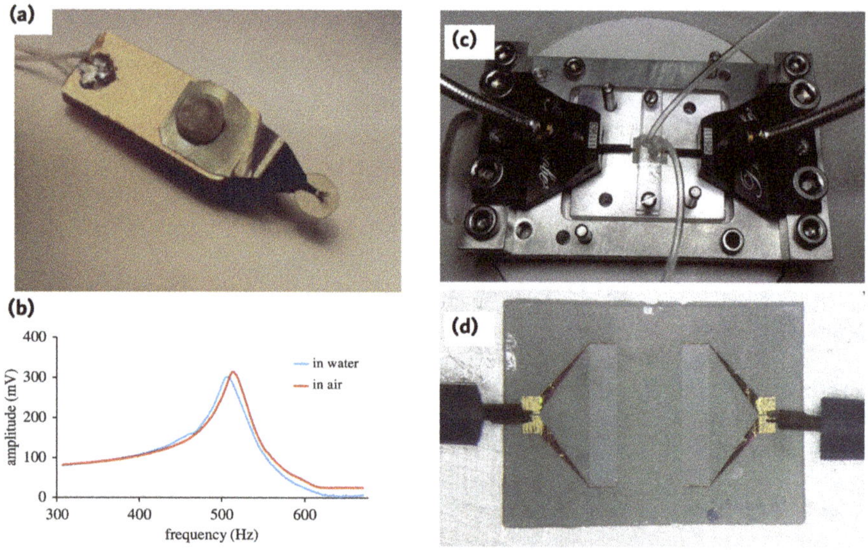

Figure 4.5. (a), (b) Photographs of a SAW biosensor and microwave probes [69]; and (c), (d) disc holder and the piezo disc mechanical oscillations [70]. (a), (b) Reprinted from [69], Copyright 2015, with permission from Elsevier. (c), (d) Reproduced from [70]. CC BY 4.0.

performed label-free and rapid detection of influenza A viruses. The limit of detection was less than 10^5 virus particles per ml. They also reported the selective, sensitive, and label-free detection of influenza viruses. The PTZ disc and the resonance peaks of the piezo disc in water and air are presented in figures 4.5(c) and (d) [70].

4.3.3 SARS virus

In December 2019, an acute virulent disease that caused severe respiratory trouble broke out in Wuhan, and the spread of cases was attributed to the transmission of a novel coronavirus, named severe acute respiratory syndrome coronavirus 2019 (SARS-CoV-2019) [71]. Most patients with SARS-CoV-2019 experience a dry cough, chest tightness, fever with or without chills, shortness of breath, and infiltration of the lungs increasing from patchy to total. Therefore, the outbreak represents a newly emerging viral sickness caused by strains that infect humans [16].

Zuo *et al* established a piezoelectric biosensor for the detection of SARS-CoV from mucus in the gas phase. This study has very important experimental results. It proved the suitability and simplicity of the piezoelectric biosensor for sensing SARS-CoV. Frequency shifts were seen to be linear to the antigen concentration in the range of 0.6–4 $\mu g\ mL^{-1}$. This biosensor has been shown perfect reproducibility nearly 100 times without a noticeable decrease in activity. The biosensor was also stable for more than 2 months. The biosensor also exhibited a short analysis time of less than than 2 min. In this study, it was reported that the difference between mucus with and without SARS-CoV was clear enough to be utilized in the detection of SARS [59].

The epidemic of SARS-CoV-2 has been declared a pandemic by the World Health Organization (WHO). After the structure of the spike protein of SARS-CoV-2 was elucidated, various and effective strategies were developed for the detection of this new coronavirus. Given the current situation, it is desirable to develop effective and reliable piezoelectric biosensors that can detect SARS-CoV-2. A piezoelectric-based biosensor for the detection of SARS-CoV has been reported by Zuo *et al* [59], and applications of the recently suggested surface chemistry will provide quick and early SARS-CoV-2 detection [72].

4.3.4 Vaccinia virus

Vaccinia virus is a complex, large DNA virus of the opoxvirus type in the poxvirus family [73]. QCM biosensor technology has been used rapidly and in real time for vaccinia virus detection. Kleo *et al* developed a QCM biosensor combined with DNA for the detection of the vaccinia virus. A piezoelectric quartz chip was designed to be used with different immobilization strategies for vaccinia virus DNA detection. For the QCM biosensor analysis, they utilized a homemade flow cell combined with a laboratory 10 MHz, gold-coated QCM chip. Figure 4.6(a) shows the homemade flow cell and the 5 μl flow chamber above the QCM biosensor chip [74].

The quick and sensitive detection of airborne organisms has become significant for applications in the areas of public health, homeland security, and biological research [75]. A QCM biosensor for airborne vaccinia virus detection was developed

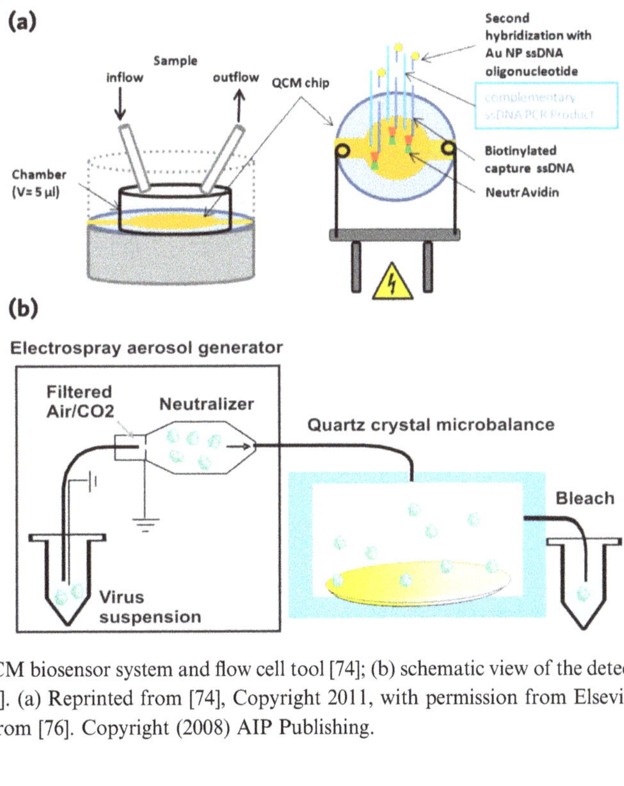

Figure 4.6. (a) QCM biosensor system and flow cell tool [74]; (b) schematic view of the detection system for the airborne virus [76]. (a) Reprinted from [74], Copyright 2011, with permission from Elsevier. (b) Reproduced with permission from [76]. Copyright (2008) AIP Publishing.

by Lee *et al.* The capture rate varied linearly with the initial virus suspension at flow rates of 2.0 and 1.1 L min^{-1}, with a concentration range of 8.5×10^8 particles mL^{-1} to 8.5×10^{10} particles mL^{-1}. Figure 4.6(b) shows a schematic of the overall experimental setup consisting of an electrospray aerosol generator and a Q-Sense model D-300 QCM biosensor electrode. This study shows the common potential for the mass-sensitivity-based detection of nanoscale biological entities in the air [76].

4.3.5 Human immunodeficiency virus

The human immunodeficiency virus (HIV) was reported as a dangerous blood-borne pathogen in the early 1980s [77]. Therefore, the identification of transmission in hospitals and the public health environment continues to be a global concern [78]. HIV type-1 glycoprotein 41 (HIV-1 gp41) is the causal factor for acquired immunodeficiency syndrome (AIDS). Very interesting methods have used the piezoelectric technique to detect HIV.

MIPs are prearranged polymers that are selective for a given target analyte. The MIP technique has been used for a broad range of applications in various fields including pharmaceutical, environmental, and medical applications, food control, and homeland security [79–81]. Lu *et al* developed a selective biosensor based on an epitope-imprinting technique for HIV type-1 (HIV-1) detection using glycoprotein 41 (gp41) [82]. A synthetic peptide with 35 amino acid residues analogous to residues 579–613 of HIV-1 gp41 was used as the template and dopamine was utilized as a functional monomer. The limit of detection for gp41 was found to be 2 ng mL^{-1}. Direct detection of gp41 was achieved using a MIP-based piezoelectric biosensor with human urine samples. Lu *et al* also reported that this biosensor was comparable to enzyme-linked immunosorbent assay analysis. The practical analytical performance of the developed biosensor was studied by evaluating the detection of gp41. Figure 4.7(a) shows the preparation procedure of the biomimetic-based QCM biosensor.

Ly *et al* developed a QCM biosensor for HIV-1 p24 detection. They used gold nanoparticles (AuNPs) to amplify a signal. AuNPs of four different sizes were

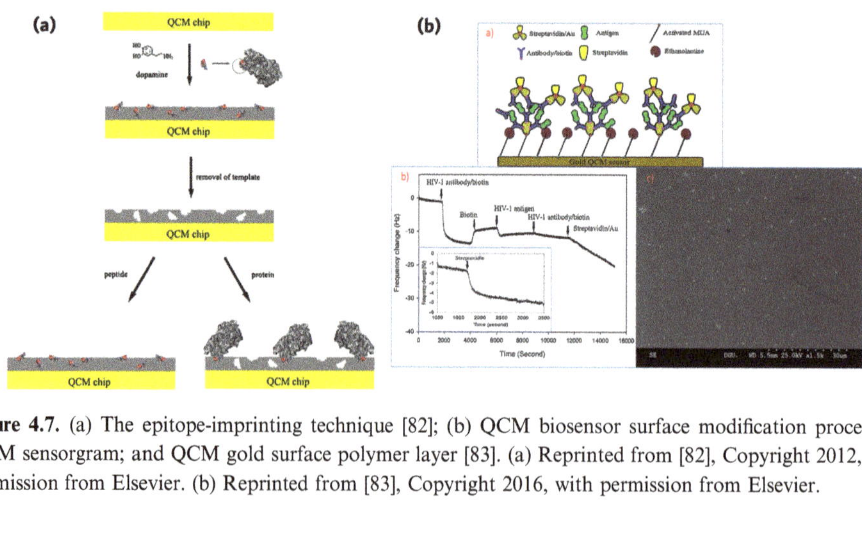

Figure 4.7. (a) The epitope-imprinting technique [82]; (b) QCM biosensor surface modification procedure; QCM sensorgram; and QCM gold surface polymer layer [83]. (a) Reprinted from [82], Copyright 2012, with permission from Elsevier. (b) Reprinted from [83], Copyright 2016, with permission from Elsevier.

synthesized. With signal-enhancing AuNPs, it was possible to identify the target at a concentration of 1 ng mL^{-1}. Figure 4.7(b) shows that exposure of the QCM biosensor surface to antibody biotin caused a decrease in frequency [83].

Minunni *et al* developed a specific RNA aptamer immobilized QCM biosensor for the protein transactivator of transcription (TAT) of HIV-1. This work utilized a specific RNA aptamer as the biological recognition element. The antibody was immobilized on a layer of carboxylated dextran previously deposited on a QCM gold electrode. The linear range obtained using the antibody was from 0 to 2.5 ppm and the limit of detection was 0.25 ppm [84]. In another study, an aptamer-based QCM biosensor for the detection of HIV-1 TAT protein was developed by Tombelli *et al.* They used a specific aptamer immobilized gold sensor surface to fabricate both a surface plasmon resonance (SPR) biosensor and a QCM biosensor. The QCM biosensor system had a linear response to HIV-1 TAT protein concentrations ranging from 0 to 1.25 ppm. They also reported that six different QCM crystals were modified to test the reproducibility of the immobilization step [84].

Marco Bisoffi *et al* fabricated a piezoelectric biosensor for the quick detection of HIV types I and II. A lithium tantalate ($LiTaO_3$)-based acoustic wave biosensor was used in the study. A portable biosensor immobilized with specific antibodies against HIV-1 and HIV-2 was developed. The quick and sensitive detection of both HIV-1 and HIV-2 was demonstrated in both human serum and cross-confounding virus. The developed biosensor detected the presence of the virus within 5 min. It also selectively distinguished between HIV-1 and HIV-2. Therefore, this piezoelectric biosensor could potentially be transformed into a tool for utilization in applications that need quick and reliable analysis of possible HIV transmission in blood donors [85].

4.3.6 Human papillomavirus

HPV is a genus of the *Papillomaviridae* family. Cervical cancer, the third most common cancer in women, can be caused by HPV. It is one of the sexually transmittable viruses. While most HPV viruses do not affect people, some types can be the reason for health problems, including genital cancer or warts [86, 87].

Fu *et al* established a quick detection technique using a combination of a piezoelectric biosensor with a gene chip for the detection of HPV. They utilized an AT-cut 10 MHz piezoelectric quartz crystal biosensor for the immobilization of single-stranded nucleic acid and subsequent hybridization reactions. Two different QCM electrodes were designed in this work. HPV6, HPV11, HPV16, and HPV18 were detected using the QCM-I electrode and only HPV18 was undetected when the QCM-II electrode was used [88].

Another paper on the topic of piezoelectric biosensors for HPV detection was written by Daniela Dell'Atti *et al.* They used label-free probes for piezoelectric biosensors based on DNA for the specific detection of HPV virus strains in human biofluids [89]. Genital HPV types are divided into subgroups, such as 6, 11, 16, 18, 33. Genital HPV is a medium-low-risk or high-risk virus often associated with invasive cervical cancer. Dell'Atti *et al* prepared a piezoelectric biosensor for the detection of HPV16 and HPV18. The linear concentration range of samples was between 50 and

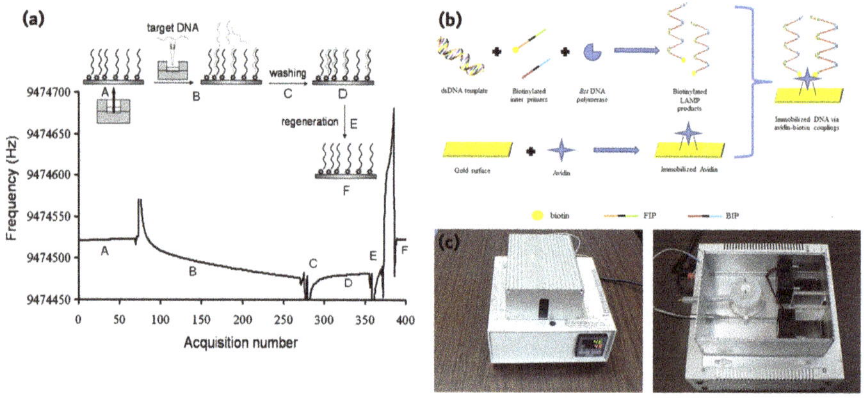

Figure 4.8. (a) Schematic diagram of a piezoelectric measurement; (b) schematic diagram of biotinylated loop-mediated isothermal amplification (LAMP) detection on a gold quartz surface; (c) the temperature control box and the QCM biosensor system. (a) Reprinted from [89], Copyright 2007, with permission from Elsevier. (b), (c) Reprinted from [90], Copyright 2013, with permission from Elsevier.

300nm and the limit of detection was calculated to be 30 nM. Figure 4.8(a) shows a classic QCM sensorgram recorded during a hybridization analysis by a quartz crystal biosensor that detects each step of the hybridization event.

Prakrankamanant *et al* developed a DNA-based QCM biosensor for the detection of HPV DNA type 58 (HPV58) (figure 4.8(b)). AT-cut 9 MHz QCM quartz crystals were utilized with a gold-coated electrode. The QCM chamber with a homemade temperature control box housing is shown in figure 4.8(c). The LAMP-QCM biosensor was exhibited better performance than traditional LAMP detection, with 100% sensitivity and 90.5% specificity. It has also been reported that the LAMP-QCM integrated biosensor system has an improved detection limit of up to ten times that of traditional LAMP [90].

4.3.7 Other viruses

African swine fever (ASF) is a highly contagious viral disease. African swine fever virus (ASFV) causes high morbidity and mortality with associated economic losses [91]. A diagnostic method is also needed for disease recognition in the early stages of infection. A quick, selective, and sensitive technique for the detection of these viruses would help to reduce the risk of epidemics. E. Uttenthaler *et al* developed a piezoelectric biosensor for direct detection of the African swine fever virus using a flow injection analysis system. The peptide-specific monoclonal antibody 18BG3 and the virus protein 73 (VP73) were utilized. The accumulation of 18BG3 and VP73 on the chip surface of this QCM biosensor caused a shift in the resonant frequency. It has been reported that different immobilization methods are suitable for coating the monoclonal antibody 18BG3. The limits of detection were calculated to be 1.0 and 0.2 μg mL^{-1} for VP73 and 18BG3 in diluted pig serum [92].

Dengue virus is a newly emerging disease and poses a serious global problem. The clinical symptoms of dengue hemorrhagic fever and dengue fever are atypical.

There are still no effective vaccines or medicines available to prevent or cure diseases caused by the dengue virus. A short response time, high accuracy, and a labor-free process are crucial to the early detection of dengue fever. The epitope approach to the synthesis of MIP film on a biosensor chip surface was described by Dar-Fu Tai *et al.* A MIP film was synthesized on a QCM biosensor chip in the presence of a pentadecapeptide. A 15-mer peptide was used as the linear epitope of the NS1 protein dengue virus [93].

The early symptoms of Ebola are endemic malaria and other febrile illnesses. It is a grave concern that that the presence of the Ebola virus cannot be detected quickly. Baca *et al* designed a SAW biosensor for Ebola virus detection. The limit of detection reached 1.9×10^4 PFU mL^{-1} before virus inactivation occurred. They reported a predicted very high sensitivity for intact, infectious Ebola virus. As part of a point-of-care methodology, the SAW biosensor device greatly facilitates infection control and rapid detection of the Ebola virus. It also works on disposable batteries [94].

Jenik *et al* designed a MIP film-based QCM biosensor for the detection of human rhinovirus (HRV) and the foot-and-mouth disease virus (FMDV), which are two agents of picornaviruses. A surface imprinting technique led to patterned polyurethane layers that replicate the shape and size properties of the target virus. The surface imprinting for HRV was confirmed using an atomic force microscope. The detection limit of the QCM biosensor was calculated to be $\sim$100 μg mL^{-1}. The selectivity of the MIP-based QCM biosensor to various HRV serotypes was evaluated using HRV2, HRV1A, HRV14, and HRV16 template molecules [95].

4.4 Conclusions

Piezoelectric-based biosensors are an excellent and practical platform for monitoring affinity interactions; their use means that analysis can be designed to be label-free, meaning that the interaction between the biosensor chip's metal surface and the target analyte is direct, and does not require the application of any specific reagent or sample processing. Quartz is a material with piezoelectric features, which means it has the ability to create a potential when mechanically squeezed or mechanically deformed when an oriented dipole is present on its crystal surface. Piezoelectric-based biosensors are extremely sensitive to mass and can detect mass changes at the nanogram level. QCMs appear to be the most common sensor platform in the quartz type of biosensor. Piezoelectric biosensors are ultrasensitive to mass changes on the surface of the crystal chip; their ability to detect a target analyte has many ideal qualities when used with specially designed nanomaterials.

When the developments in quick and early COVID-19 detection are considered, it appears that the development of biosensors is a very important step in monitoring currently known and future virus threats. Biosensors that can further increase detection sensitivity can be developed and used for early virus detection. As seen in the past year, the spread of the severe acute respiratory syndrome coronavirus has had a major impact on the economy and societies, affecting and changing the lives of people worldwide. As a result, the development of portable and wearable biosensors that can continuously monitor viruses has become an important area of research. For this reason, piezoelectric biosensors will become tools for further research in this area.

References and further reading

[1] Müller P and Schmid M 2019 Intelligent packaging in the food sector: a brief overview *Foods* **8** 16

[2] Stetter J R, Penrose W R and Yao S 2003 Sensors, chemical sensors, electrochemical sensors, and ECS *J. Electrochem. Soc.* **150** S11

[3] Manoj D, Shanmugasundaram S and Anandharamakrishnan C 2021 Nanosensing and nanobiosensing: concepts, methods, and applications for quality evaluation of liquid foods *Food Control* **126** 108017

[4] Saylan Y, Akgönüllü S, Yavuz H, Ünal S and Denizli A 2019 Molecularly imprinted polymer based sensors for medical applications *Sensors (Switzerland)* **19** 1279

[5] Velasco-Garcia M N and Mottram T 2003 Biosensor technology addressing agricultural problems *Biosyst. Eng.* **84** 1–12

[6] Rezaei Z and Mahmoudifard M 2019 Pivotal role of electrospun nanofibers in microfluidic diagnostic systems-a review *J. Mater. Chem.* B **7** 4602–19

[7] Mohammed M I and Desmulliez M P Y 2011 Lab-on-a-chip based immunosensor principles and technologies for the detection of cardiac biomarkers: a review *Lab Chip* **11** 569–95

[8] Qureshi A, Gurbuz Y and Niazi J H 2012 Biosensors for cardiac biomarkers detection: a review *Sensors Actuators* B **171–2** 62–76

[9] Tibuzzi A, Rea G, Pezzotti G, Esposito D, Johanningmeier U and Giardi M T 2007 A new miniaturized multiarray biosensor system for fluorescence detection *J. Phys. Condens. Matter* **19** 395006

[10] Akgönüllü S, Yavuz H and Denizli A 2020 SPR nanosensor based on molecularly imprinted polymer film with gold nanoparticles for sensitive detection of aflatoxin B1 *Talanta* **219** 121219

[11] Rahtuvanoğlu A, Akgönüllü S, Karacan S and Denizli A 2020 Biomimetic nanoparticles based surface plasmon resonance biosensors for histamine detection in foods *ChemistrySelect* **5** 5683–92

[12] Heller D A, Baik S, Eurell T E and Strano M S 2005 Single-walled carbon nanotube spectroscopy in live cells: towards long-term labels and optical sensors *Adv. Mater.* **17** 2793–9

[13] Lyu Y, Cui D, Huang J, Fan W, Miao Y and Pu K 2019 Near-infrared afterglow semiconducting nano-polycomplexes for the multiplex differentiation of cancer exosomes *Angew. Chemie* **131** 5037–41

[14] Ren X, Wei J, Ren J, Qiang L, Tang F and Meng X 2015 A sensitive biosensor for the fluorescence detection of the acetylcholinesterase reaction system based on carbon dots *Colloids Surf.* B **125** 90–5

[15] Li X, Yang L, Li X, Yan S, Ren Y, Wang M, Liu P, Dong Y and Zhang C 2016 $[Cu(phen)2]^{2+}$ acts as electrochemical indicator and anchor to immobilize probe DNA in electrochemical DNA biosensor *Anal. Biochem.* **492** 56–62

[16] Narita F, Wang Z, Kurita H, Li Z, Shi Y, Jia Y and Soutis C 2021 A review of piezoelectric and magnetostrictive biosensor materials for detection of COVID-19 and other viruses *Adv. Mater.* **33** 2005448

[17] Chen Y, Qian C, Liu C, Shen H, Wang Z, Ping J, Wu J and Chen H 2020 Nucleic acid amplification free biosensors for pathogen detection *Biosens. Bioelectron.* **153** 112049

[18] Pan Y, Zhang L, Cao B, Xue X, Liu W, Zhang C and Wang W 2020 Effects of temperature and humidity on the performance of a PECH polymer coated SAW sensor *RSC Adv.* **10** 18099–106

[19] Jeng M J, Sharma M, Li Y C, Lu Y C, Yu C Y, Tsai C L, Huang S F, Chang L B and Lai C S 2020 Surface acoustic wave sensor for c-reactive protein detection *Sensors (Switzerland)* **20** 1–11

[20] Temel F 2020 One novel calix[4]arene based QCM sensor for sensitive, selective and high performance-sensing of formaldehyde at room temperature *Talanta* **211** 120725

[21] Haghighi E and Zeinali S 2020 Formaldehyde detection using quartz crystal microbalance (QCM) nanosensor coated by nanoporous MIL-101(Cr) film *Microporous Mesoporous Mater.* **300** 110065

[22] Jin X, Zhang Y P, Li D M, Ma D, Zheng S R, Wu C H, Li J Y and Zhang W G 2020 The interaction of an amorphous metal-organic cage-based solid (aMOC) with miRNA/DNA and its application on a quartz crystal microbalance (QCM) sensor *Chem. Commun.* **56** 591–4

[23] Mujahid A, Afzal A, Glanzing G, Leidl A, Lieberzeit P A and Dickert F L 2010 Imprinted sol-gel materials for monitoring degradation products in automotive oils by shear transverse wave *Anal. Chim. Acta* **675** 53–7

[24] Afzal A, Mujahid A, Schirhagl R, Bajwa S Z, Latif U and Feroz S 2017 Gravimetric viral diagnostics: QCM based biosensors for early detection of viruses *Chemosensors* **5** 1–25

[25] Saylan Y, Erdem Ö, Ünal S and Denizli A 2019 An alternative medical diagnosis method: biosensors for virus detection *Biosensors* **9** 65

[26] Suleiman A A and Guilbault G G 1994 Recent developments in piezoelectric immunosensors. A review *Analyst* **119** 2279–82

[27] King W H 1964 Piezoelectric sorption detector *Anal. Chem.* **36** 1735–9

[28] Suleiman A A and Guilbault G G 1991 Piezoelectric (PZ) immunosensors and their applications *Anal. Lett.* **24** 1283–92

[29] Kurosawa S, Kitajima H, Ogawa Y, Muratsugu M, Nemoto E and Kamo N 1993 Resonant frequency of a piezoelectric quartz crystal in contact with solutions *Anal. Chim. Acta* **274** 209–17

[30] Han D H and Kang L H 2020 Piezoelectric properties of paint sensor according to piezoelectric materials *Funct. Compos. Struct.* **2** 2–13

[31] Shindo Y, Domon W and Narita F 1998 Dynamic bending of a symmetric piezoelectric laminated plate with a through crack *Theor. Appl. Fract. Mech.* **28** 175–82

[32] Fay B, Lewin P A, Ludwig G, Sessler G M and Yang G 1992 The influence of spatial polarization distribution on spot poled PVDF membrane hydrophone performance *Ultrasound Med. Biol.* **18** 625–35

[33] Yoshimura H N, Molisani A L, Narita N E, Manholetti J L A and Cavenaghi J M 2006 Mechanical properties and microstructure of zinc oxide varistor ceramics *Mater. Sci. Forum* **530–1** 408–13

[34] IEEE 2012 1983 IEEE ultrasonics symposium abstracts *IEEE Trans. Sonics Ultrason.* **32** 78–122

[35] Kamiya T 1996 Calculation of crystal structures, dielectric constants and piezoelectric properties of wurtzite-type crystals using ab-initio periodic Hartree-Fock method *Jpn. J. Appl. Phys.* **35** 4421

[36] Narita F and Shindo Y 2015 Piezoelectric detection and response characteristics of barium titanate unimorph cantilevers under AC electric fields *Int. J. Metall. Mater. Eng.* **1** 10–3

[37] Marutake M 1956 A calculation of physical constants of ceramic barium titanate *J. Phys. Soc. Jpn.* **11** 807–14

[38] Hwa L G, Lu C L and Liu L C 2000 Elastic moduli of calcium alumino-silicate glasses studied by Brillouin scattering *Mater. Res. Bull.* **35** 1285–92

[39] Carr P H 1967 Measurement of the piezoelectric constant of quartz at gigacycle frequencies *J. Acoust. Soc. Am.* **41** 75–83

[40] Kuma S and Woldemariam M M 2019 Structural, electronic, lattice dynamic, and elastic properties of $SnTiO_3$ and $PbTiO_3$ using density functional theory *Adv. Condens. Matter Phys.* **2019** 3176148

[41] Haun M J, Furman E, Jang S J, McKinstry H A and Cross L E 2006 Thermodynamic theory of the lead zirconate-titanate solid solution system, part I: phenomenology ferroelectrics **99** 13–25

[42] Ravina R, Dalal A, Mohan H, Prasad M and Pundir C S 2020 Detection methods for influenza A H1N1 virus with special reference to biosensors: a review *Biosci. Rep.* **40** 1–18

[43] Perumal V and Hashim U 2014 Advances in biosensors: principle, architecture and applications *J. Appl. Biomed.* **12** 1–15

[44] Sullivan C K O and Guilbault G G 1999 Commercial quartz crystal microbalances—theory and applications *Biosens. Bioelectron.* **14** 663–70

[45] Anderson H, Jönsson M, Vestling L, Lindberg U and Aastrup T 2007 Quartz crystal microbalance sensor design. I. Experimental study of sensor response and performance *Sensors Actuators* B **123** 27–34

[46] Pan M, Li R, Xu L, Yang J, Cui X and Wang S 2018 Reproducible molecularly imprinted piezoelectric sensor for accurate and sensitive detection of ractopamine in swine and feed products *Sensors (Switzerland)* **18** 1870

[47] Akgönüllü S, Battal D, Yalcin M S, Yavuz H and Denizli A 2019 Rapid and sensitive detection of synthetic cannabinoids JWH-018, JWH-073 and their metabolites using molecularly imprinted polymer-coated QCM nanosensor in artificial saliva *Microchem. J.* **153** 104454

[48] Battal D, Akgönüllü S, Yalcin M S, Yavuz H and Denizli A 2018 Molecularly imprinted polymer based quartz crystal microbalance sensor system for sensitive and label-free detection of synthetic cannabinoids in urine *Biosens. Bioelectron.* **111** 10–7

[49] Lee L T, Mohamed M A, Yahya I, Kulothungan J, Muruganathan M and Mizuta H 2018 Comparison of piezoelectric energy harvesting performance using silicon and graphene cantilever beam *Microsyst. Technol.* **24** 3783–9

[50] Meléndrez D, Hampitak P, Jowitt T, Iliut M and Vijayaraghavan A 2021 Development of an open-source thermally stabilized quartz crystal microbalance instrument for biomolecule-substrate binding assays on gold and graphene *Anal. Chim. Acta* **1156**

[51] Lee M H, Thomas J L, Tseng H Y, Lin W C, Liu B D and Lin H Y 2011 Sensing of digestive proteins in saliva with a molecularly imprinted poly(ethylene-co-vinyl alcohol) thin film coated quartz crystal microbalance sensor *ACS Appl. Mater. Interfaces* **3** 3064–71

[52] Zhao X, He Y, Wang Y, Wang S and Wang J 2020 Hollow molecularly imprinted polymer based quartz crystal microbalance sensor for rapid detection of methimazole in food samples *Food Chem.* **309** 125787

[53] Cervera-Chiner L, Juan-Borrás M, March C, Arnau A, Escriche I, Montoya Á and Jiménez Y 2018 High fundamental frequency quartz crystal microbalance (HFF-QCM) immunosensor for pesticide detection in honey *Food Control* **92** 1–6

[54] Li D-M, Zhang S-Y, Wan J-Y, Zhang W-Q, Yan Y-L, Tang X-H, Zheng S-R, Cai S and Zhang W-G 2021 A new hydrazone-linked covalent organic framework for Fe(III) detection by fluorescence and QCM technologies *CrystEngComm*

[55] Cervera-Chiner L, Jiménez Y, Montoya Á, Juan-Borrás M, Pascual N, Arnau A and Escriche I 2020 High fundamental frequency quartz crystal microbalance (HFF-QCMD) immunosensor for detection of sulfathiazole in honey *Food Control* **115** 107296

[56] Su L, Zou L, Fong C C, Wong W L, Wei F, Wong K Y, Wu R S S and Yang M 2013 Detection of cancer biomarkers by piezoelectric biosensor using PZT ceramic resonator as the transducer *Biosens. Bioelectron.* **46** 155–61

[57] Buchatip S, Ananthanawat C, Sithigorngul P, Sangvanich P, Rengpipat S and Hoven V P 2010 Detection of the shrimp pathogenic bacteria, *Vibrio harveyi*, by a quartz crystal microbalance-specific antibody based sensor *Sensors Actuators* B **145** 259–64

[58] Li B, Hu X, Zhang Q, Peng X and Xiang Y 2021 Improved piezoelectricity of polylactide using vitamin B2 for poling-free mechanical and acoustic nanogenerators *J. Mater. Sci.* **56** 902–12

[59] Zuo B, Li S, Guo Z, Zhang J and Chen C 2004 Piezoelectric immunosensor for SARS-associated coronavirus in sputum *Anal. Chem.* **76** 3536–40

[60] Kreutzberg K 1991 Hepatitis B virus *Fortschr. Med.* **109** 48

[61] Xu T, Miao J, Wang Z, Yu L and Li C M 2011 Micro-piezoelectric immunoassay chip for simultaneous detection of Hepatitis B virus and α-fetoprotein *Sensors Actuators* B **151** 370–6

[62] Zhou X, Liu L, Hu M, Wang L and Hu J 2002 Detection of hepatitis B virus by piezoelectric biosensor *J. Pharm. Biomed. Anal.* **27** 341–5

[63] Patel J and Sharma P 2020 Design of a novel rapid immunoassay for simultaneous detection of hepatitis C virus core antigen and antibodies *Arch. Virol* **165** 627–41

[64] Skládal P, Dos Santos Riccardi C, Yamanaka H and Da Costa P I 2004 Piezoelectric biosensors for real-time monitoring of hybridization and detection of hepatitis C virus *J. Virol. Methods* **117** 145–51

[65] Webster R G, Peiris M, Chen H and Guan Y 2006 H5N1 outbreaks and enzootic influenza *Biodiversity* **7** 51–5

[66] Gamblin S J and Skehel J J 2010 Influenza hemagglutinin and neuraminidase membrane glycoproteins *J. Biol. Chem.* **285** 28403–9

[67] Wangchareansak T, Thitithanyanont A, Chuakheaw D, Gleeson M P, Lieberzeit P A and Sangma C 2014 A novel approach to identify molecular binding to the influenza virus H5N1: screening using molecularly imprinted polymers (MIPs) *MedChemComm* **5** 617–21

[68] Li D, Wang J, Wang R, Li Y, Abi-Ghanem D, Berghman L, Hargis B and Lu H 2011 A nanobeads amplified QCM immunosensor for the detection of avian influenza virus H5N1 *Biosens. Bioelectron.* **26** 4146–54

[69] Jiang Y, Tan C Y, Tan S Y, Wong M S F, Chen Y F, Zhang L, Yao K, Gan S K E, Verma C and Tan Y J 2015 SAW sensor for Influenza A virus detection enabled with efficient surface functionalization *Sensors Actuators* B **209** 78–84

[70] Erofeev A S, Gorelkin P V, Kolesov D V, Kiselev G A, Dubrovin E V and Yaminsky I V 2019 Label-free sensitive detection of influenza virus using PZT discs with a synthetic sialylglycopolymer receptor layer *R. Soc. Open Sci.* **6** 1–13

[71] Xiao S Y, Wu Y and Liu H 2020 Evolving status of the 2019 novel coronavirus infection: proposal of conventional serologic assays for disease diagnosis and infection monitoring *J. Med. Virol.* **92** 464–7
[72] Pandey L M 2020 Design of engineered surfaces for prospective detection of Sars-CoV-2 using quartz crystal microbalance based techniques *Expert Rev. Proteomics* **17** 425–32
[73] Chung C-S, Chen C-H, Ho M-Y, Huang C-Y, Liao C-L and Chang W 2006 Vaccinia virus proteome: identification of proteins in vaccinia virus intracellular mature virion particles *J. Virol.* **80** 2127–40
[74] Kleo K, Kapp A, Ascher L and Lisdat F 2011 Detection of vaccinia virus DNA by quartz crystal microbalance *Anal. Biochem.* **418** 260–6
[75] Zeng Q Y, Westermark S O, Rasmuson-Lestander Å and Wang X R 2006 Detection and quantification of Cladosporium in aerosols by real-time PCR *J. Environ. Monit.* **8** 153–60
[76] Lee J, Jang J, Akin D, Savran C A and Bashir R 2008 Real-time detection of airborne viruses on a mass-sensitive device *Appl. Phys. Lett.* **93** 8–11
[77] Malm K, Von Sydow M and Andersson S 2009 Performance of three automated fourth-generation combined HIV antigen/antibody assays in large-scale screening of blood donors and clinical samples *Transfus. Med.* **19** 78–88
[78] Weber B, Berger A, Rabenau H and Doerr H W 2002 Evaluation of a new combined antigen and antibody human immunodeficiency virus screening assay, VIDAS HIV DUO Ultra *J. Clin. Microbiol.* **40** 1420–6
[79] Bakhshpour M, Göktürk I, Bereli N and Denizli A 2020 Molecularly imprinted cryogel cartridges for the selective recognition of tyrosine *Biotechnol. Prog.* **36** 3006
[80] Öngün E, Akgönüllü S, Yavuz H and Denizli A 2020 Synthesis of molecularly imprinted magnetic nanoparticles for selective cytidine adsorption *Sep. Sci. Plus* **4** 147–56
[81] Esentürk M K, Akgönüllü S, Yılmaz F and Denizli A 2019 Molecularly imprinted based surface plasmon resonance nanosensors for microalbumin detection *J. Biomater. Sci. Polym. Ed.* **30** 646–61
[82] Lu C H, Zhang Y, Tang S F, Fang Z B, Yang H H, Chen X and Chen G N 2012 Sensing HIV related protein using epitope imprinted hydrophilic polymer coated quartz crystal microbalance *Biosens. Bioelectron.* **31** 439–44
[83] Ly T N, Park S and Park S J 2016 Detection of HIV-1 antigen by quartz crystal microbalance using gold nanoparticles *Sensors Actuators* B **237** 452–8
[84] Minunni M, Tombelli S, Gullotto A, Luzi E and Mascini M 2004 Development of biosensors with aptamers as bio-recognition element: the case of HIV-1 Tat protein *Biosens. Bioelectron.* **20** 1149–56
[85] Bisoffi M, Severns V, Branch D W, Edwards T L and Larson R S 2013 Rapid detection of human immunodeficiency virus types 1 and 2 by use of an improved piezoelectric biosensor *J. Clin. Microbiol.* **51** 1685–91
[86] Mahmoodi P, Fani M, Rezayi M, Avan A, Pasdar Z, Karimi E, Amiri I S and Ghayour-Mobarhan M 2019 Early detection of cervical cancer based on high-risk HPV DNA-based genosensors: a systematic review *BioFactors* **45** 101–17
[87] Vinodhini K, Shanmughapriya S, Das B C and Natarajaseenivasan K 2012 Prevalence and risk factors of HPV infection among women from various provinces of the world *Arch. Gynecol. Obstet.* **285** 771–7

[88] Genesensors C, Fu W, Huang Q, Wang J, Liu M, Huang J and Chen B 2004 Sensors & transducers detection of human papilloma virus with piezoelectric quartz *Sensors (Peterborough, NH)* **42** 214–9

[89] Dell'Atti D, Zavaglia M, Tombelli S, Bertacca G, Cavazzana A O, Bevilacqua G, Minunni M and Mascini M 2007 Development of combined DNA-based piezoelectric biosensors for the simultaneous detection and genotyping of high risk Human Papilloma Virus strains *Clin. Chim. Acta* **383** 140–6

[90] Prakrankamanant P, Leelayuwat C, Promptmas C, Limpaiboon T, Wanram S, Prasongdee P, Pientong C, Daduang J and Jearanaikoon P 2013 The development of DNA-based quartz crystal microbalance integrated with isothermal DNA amplification system for human papillomavirus type 58 detection *Biosens. Bioelectron.* **40** 252–7

[91] Galindo I and Alonso C 2017 African swine fever virus: a review *Viruses* **9** 103

[92] Uttenthaler E, Kößlinger C and Drost S 1998 Characterization of immobilization methods for African swine fever virus protein and antibodies with a piezoelectric immunosensor *Biosens. Bioelectron.* **13** 1279–86

[93] Tai D, Lin C, Wu T and Chen L 2005 Recognition of dengue virus protein using epitope-mediated molecularly imprinted film of a pentadecapeptide onto a quartz crystal microbalance *Anal. Chem.* **77** 5140–3

[94] Baca J T, Severns V, Lovato D, Branch D W and Larson R S 2015 Rapid detection of Ebola virus with a reagent-free, point-of-care biosensor *Sensors (Switzerland)* **15** 8605–14

[95] Jenik M, Schirhagl R, Schirk C, Hayden O, Lieberzeit P, Blaas D, Paul G and Dickert F L 2009 Sensing picornaviruses using molecular imprinting techniques on a quartz crystal microbalance *Anal. Chem.* **81** 5320–6

IOP Publishing

Biosensors for Virus Detection

Adil Denizli and Yeşeren Saylan

Chapter 5

Fluorescence biosensors for virus detection

Cem Esen and Sudhirkumar Shinde

As the world has been struggling with the COVID-19 pandemic since the beginning of 2020, it is well understood that rapid and reliable virus detection is of paramount importance for public health. In this context, fluorescence biosensors are one of the gold standard detection methods used for the analysis of viral infections. This chapter provides fundamental knowledge about the use of fluorescence biosensors for virus detection and the latest findings in the field of fluorescence biosensors, especially those that incorporate molecularly imprinted polymers as state-of-the-art materials, which pave the way for the detection of a wide variety of viruses. Herein, advances in the use of fluorescence-based biosensors for the detection of viruses are discussed in the light of recent papers, along with future perspectives.

5.1 Introduction

Infectious diseases, in particular, the current pandemic caused by coronavirus disease 2019 (COVID-19), pose an omnipresent threat to public health. Globally, as of 9 June 2021, there have been 173 609 772 confirmed cases of COVID-19, including 3 742 653 deaths, reported to the World Health Organization (WHO) [1]. It is likely that the main reason for the high prevalence of virus disease is the lack of effective point-of-care detection methods [2].

Viruses are obligate intracellular parasites that require a host cell in order to propagate and to establish genetic material replication. Furthermore, viruses constantly change their genetic makeup as a way of evading the host's immune system; potentially, they can cause major diseases and even death [3–5]. This is why the early detection of viruses is of paramount importance for point-of-care diagnostics [6]. Polymerase chain reactions [7], enzyme-linked immunosorbent assays [8], and reverse transcription-polymerase chain reactions [9] are widely used to detect viruses. In this context, biosensors are also vital for virus detection.

doi:10.1088/978-0-7503-3867-7ch5

© IOP Publishing Ltd 2021

5.1.1 Biosensors

Since the design of glucose oxidase biosensors introduced by Clark and Lyons in 1962, biosensors have been employed to detect various analytes, including viruses. Biosensors are analytical instruments consisting of an analyte, a bioreceptor, a transducer, and a measurable signal [10]. Following the capture of an analyte, biological responses are converted into signals. Based on the transduction principle involved, biosensors are mainly classified into three classes: optical [11], electrochemical [12], and piezoelectric [13].

Optical transducers are the simplest type of biosensor, which detect the analyte by measuring changes in the fluorescence, absorption, or reflectance of the sensing material. This technique simply exploits the color change of the sensing materials when their size or concentration changes due to interaction with the analyte [13]. The second class of biosensor is electrochemical biosensors, which can be classified as amperometric, potentiometric, impedimetric, or conductometric. Finally, the third class of biosensor comprises the micromechanical transducers; the basis of this technique is the measurement of changes in the resonance frequency. Quartz crystal microbalances (QCMs) with piezoelectric properties take advantage of frequency changes or the deflection of the sensing material [13]. Furthermore, it should be noted that in addition to the abovementioned three types of biosensor, other devices, such as plasmonic [14], microfluidic [15], cantilever-based [16], and field-effect transistor-based electronic [17] biosensors have been developed. Biosensors have turned out to be important for the detection of various kinds of virus over the last two decades.

In this context, optical biosensors serve as an alternative method for virus detection owing to their safe, straightforward use and cost-effective technology, which does not require nucleic acid amplification [18]. Fluorescence, surface plasmons, surface-enhanced Raman scattering (SERS), and colorimetric techniques have all been used for the detection of viruses, such as human immunodeficiency virus (HIV), hepatitis, Ebola virus, Zika virus, norovirus, dengue, influenza virus, and so on. The sensitivity of a viral detection method can be enormously improved by using materials that exhibit surface plasmons or SERS, but this often requires advanced instrumentation for detection. Although fluorescence lacks high sensitivity, it is a promising approach for point-of-care diagnostics, due to its relatively less complicated instrumentation, ease of use, lower cost, and the absence of a nucleic acid amplification requirement [19]. Herein, fluorescence biosensors and their applications in virus detection are presented in the light of recent literature, and their future outlook and perspectives are also discussed.

Viral biosensors can be categorized into different types, depending upon their different recognition units: immuno-, DNA-, antigen-, cell-, and molecular imprinting-based biosensors (figure 5.1). In immunosensors, virus-specific antibodies, DNA aptamers, or single-stranded oligonucleic acids (ssDNA or ssRNA) are used to detect selective viruses, virus antigens, or virus DNA with high affinity and specificity. Cell-based biosensing, meaning the collection of information about cytopathic effects, is used to examine viral diseases. In this biosensor domain, synthetic recognition units such as molecularly imprinted polymers (MIPs) are used to detect target viral antigens/antibodies with comparable affinity and selectivity.

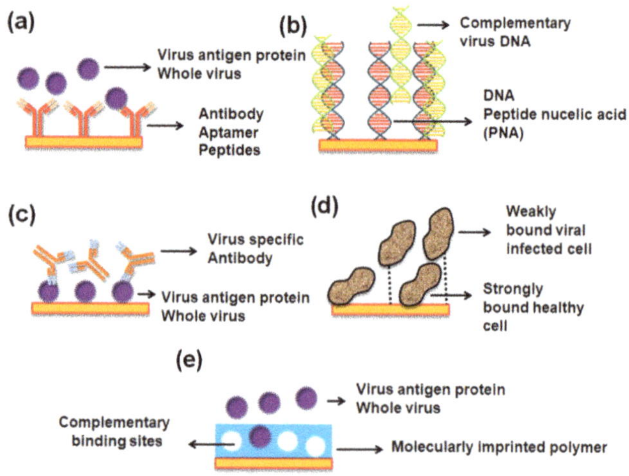

Figure 5.1. Different types of recognition-agent-based biosensor and virus target: (a) immunosensor (or antibody-based biosensor), (b) DNA-based biosensor, (c) antigen-based biosensor, (d) cell-based biosensor, and (e) molecular imprinting-based biosensor. Reprinted by permission from Springer (GmbH): Springer Nature, Communications Biology [20], Copyright (2021) CC-BY 4.0.

5.2 Fluorescence biosensors

Fluorescence-based optical biosensors are the single largest group of sensors at present, due to the commercial availability of numerous fluorescent probes, high-quality optical fibers, and suitable optical instruments [19, 21, 22]. Moreover, the absence of a requirement for high excitation power for fluorescence makes it a cost-effective, cheap, and readily available tool. Fluorescence biosensors have different parameters, such as intensity, energy transfer, lifetime, and quantum yield, which can be used for virus detection [19, 23].

Förster resonance energy transfer (FRET) is often employed in fluorescent biosensors to detect close interactions (<10 nm) between an analyte and a fluorophore. FRET is a mechanism in which incident radiation is absorbed and non-radiatively transferred from a donor to an acceptor using long-range dipole–dipole coupling [19, 24]. Based on the excitation and emission wavelengths, the fluorescence emission of sensors can be classified as upconverting or downconverting. When the emitted wavelength is shorter than the excitation wavelength (anti-Stokes shift), upconversion happens. On the other hand, downconversion is the more common mode of linear fluorescence, which uses a second near-infrared window (1.0–1.7 μm) by emitting longer wavelengths than the excitation wavelengths [19, 25].

5.2.1 Latest applications of fluorescence biosensors for virus detection

Fluorescence spectroscopy is one of the spectroscopic techniques that has been explored for virus detection at ultratrace level. The sensitivity of this method can be improved by using strong fluorescent molecules. The fluorescence-based detection methods for viruses are summarized in table 5.1. Plasmon-enhanced fluorescence (PEF) is a technique that provides better emission enhancement with decreasing lifetimes. This has allowed researchers to expand the field of fluorescence by incorporating weak quantum emitters, which avoid the opportunity to perform imaging at high resolutions. Metal-enhanced fluorescence (MEF) or plasmon-enhanced fluorescence, also called surface enhanced fluorescence (SEF), is the phenomenon of increasing the fluorescence intensity of a fluorophore material using a plasmonic nanomaterial (usually a metal). The combination of metallic nanostructure and fluorophore electrons with close proximity is associated local plasmonic electric field. Enhanced emission meaning enhanced fluorescence intensity results from the increased electric field of fluorophore [26, 27]. This phenomenon is schematically presented in figure 5.2(a).

An important condition is the selection of a suitable fluorophore so that the optical absorption bands of the fluorophore and the metal must overlap. Another important condition is the distance among the fluorophore and plasmonic surface. If the distance is in between 1-10 nm then fluorophore can be excited by the process of non-radiative localized field of the plasmon dipole. This phenomenon is also called FRET. Recently, Shrivastav *et al* have extensively reviewed the above theory and other SEF-based experimental biosensor designs [20].

Takemaura *et al* developed a localized surface plasmon resonance (LSPR)-induced fluorescence immunosensor for fast and specific influenza virus diagnosis [29]. In this method, L-cystamine capped quaternary CdSeTeS quantum dots (QDs) were linked to the anti-hemagglutinin antibody (anti-HA Ab). Furthermore, L-cystamine functionalized AuNPs were conjugated with anti-neuraminidase antibody (anti-NA Ab). The anti-NA Ab-QDs were used for recognition of the influenza virus antigen and resulted in a change in enhanced fluorescence intensity. The detection of virus using the reported sensor was achieved within 10 min with limit of detection (LOD) of 0.03 pg/ml virus concentration.

A novel detection and quantification method for multiple serotypes of HIV (A, B, C, D, E, G, and subtype panel) was proposed by Inci *et al* [28]. The HIV sensor consisted of an amine-modified polystyrene surface on which the layered deposition of AuNPs, NeutrAvidin, and biotinylated anti-gp120 polyclonal antibody took place. Figure 5.2(b) presents a schematic of the fabricated sensor, which was able to efficiently detect subtype D from whole blood samples within 70 min

The lateral flow technique in combination with SEF (figure 5.3) was used for the detection of Ebola virus glycoprotein using multifunctional nanospheres (red nanoparticles (RNs)@Au, a combination of RN QDs and AuNPs) playing the role of signal reporter [30]. RNs@Au contains approximately hundreds of quantum dots and dozens of AuNPs, which enhances both the fluorescence and the colorimetric signal, compared to single QDs or AuNPs. The nanospheres were modified with antibody and streptavidin (known as Ab-RNs@Au-SA) and bound with the test line for the detection of virus protein. Furthermore, the signal was enhanced by biotin-modified RNs@Au along with the virus protein. The method was successfully applied for virus detection in spiked urine, plasma, and tap water samples with a fast assay time of 20 min.

Table 5.1. Different fluorescence techniques used to detect various viruses.

Method	Virus	Detection element	Detection limit	Ref.
Fluorescence	Dengue	Immunofluorescence-assay-detecting sandwich complex of antibody conjugated silica microbeads and fluorescently labeled dual antibody	1×10^4 PFU mL^{-1}	[32]
	HIV or hepatitis B	Microbeads optically barcoded by CdSeS QDs coated with capture DNA adhered to a chip	1×10^3 copies mL^{-1}	[33]
	Avian influenza H_9N_2	Antibody conjugated fluorescent nanobioprobes coupled with antibody conjugated immunomagnetic beads to create fluorescent bio-targeting bifunctional cells	8.94×10^6 fg mL^{-1}	[34]
	Influenza (H_1N_1 DNA)	A sandwich complex between a CdTe QD and a protein-binding aptamer further amplified with streptavidin	3.45×10^6 fM	[35]
Plasmon-enhanced fluorescence	Ebola	Hybrid microfluid and optofluidic device with target RNA functionalized magnetic microbeads	0.2 PFU mL^{-1}	[36]
	Avian influenza H_5N_1 (rHA protein)	Influenza aptamers immobilized on AgNPs forming a complex with thiazole orange in the presence of rHA influenza protein	3.5×10^6 fg mL^{-1}	[31]
	Influenza (H_1N_1 and H_3N_2)	Binding of antibody conjugated AuNPs and CdSeTeS QDs to virus particles	H_1N_1: 30 fg mL^{-1} (water), 4×10^2 fg mL^{-1} (human serum) H_3N_2: 10 PFU mL^{-1}	[29]
	Norovirus	Norovirus antibody attached to a biosensor chip with QD antigens	1×10^4 fg mL^{-1} (4.3×10^5 copies mL^{-1})	[37]
	SARS-CoV	Fluorescently labeled antibodies attached to AuNPs	1×10^2 fg mL^{-1}	[38]
	Infectious bronchitis virus (IBV)	IBV Antibody (Ab) modified by fluorescent dye labeling (dyed-IBV-Ab) and MoS_2(MoS_2-IBV-Ab bioconjugates)	$4.6 \times 10^{\ 2}$ egg infectious dose (EID) 50 mL^{-1}	[39]

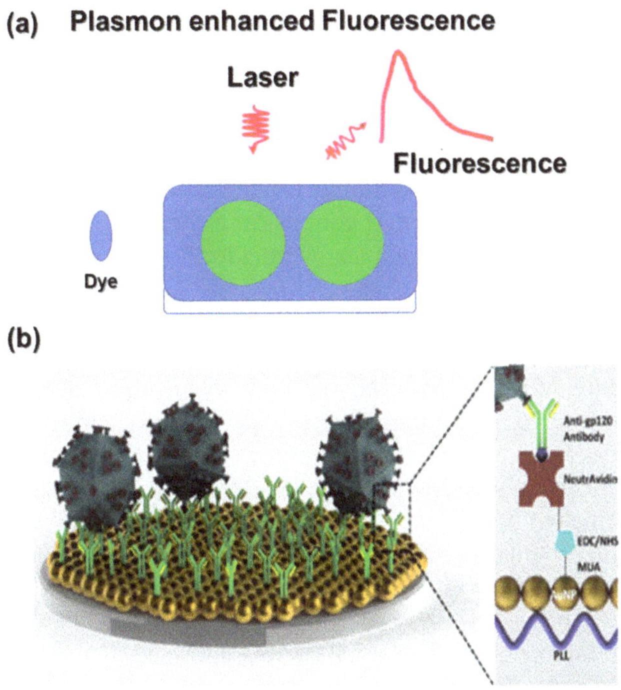

Figure 5.2. (a) Schematic of the surface-enhanced fluorescence principle. (b) HIV detection platform on gold NP array. Reprinted with permission from [28]. Copyright (2013), American Chemical Society.

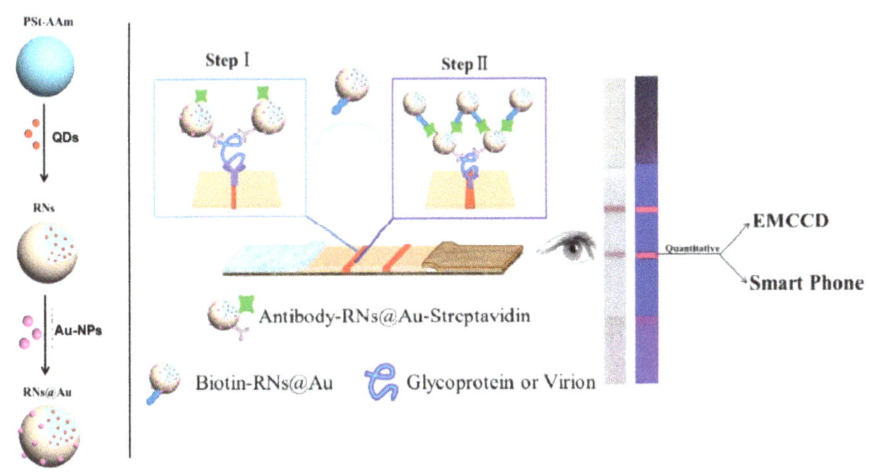

Figure 5.3. Schematic illustration of the preparation of RNs@Au (left) and schematic illustration of the two-step detection of Ebola virus using RNs@Au-based lateral flow immunoassay test strips (right). Reprinted with permission from [30]. Copyright (2017), American Chemical Society.

A guanine-enriched aptamer-based fluorescence biosensor for detection of the recombinant hemagglutinin (rHA) protein of H_5N_1 was reported, which utilized the surface of Ag@SiO_2 nanoparticles (figure 5.4(a)) and functioned as an MEF sensing platform [31]. Thiazole orange (TO) was used as a fluorescent tag that reported the G-quadruplex secondary structure induced by an aptamer-rHA binding event. In the absence of rHA protein, almost no fluorescence emission was observed with free TO in the solution (figure 5.4(b)). The aptamer strands bound with the rHA protein to form a stable G-quadruplex complex with added rHA protein. The resultant assembly binds to TO dye and excite fluorescence emission in it. Moreover, the excited-state dye with the G-quadruplex complex was in proximity with the surface of the Ag@SiO_2 nanoparticles. The metal nanoparticle surface acted as an MEF detection substrate. MEF occurs when fluorophores are close to the surface of metal nanoparticles. Owing to the coupling between fluorophores and metal nanoparticles, the detection sensitivity was significantly improved by using surface plasmon field-enhanced fluorescence. The detection of the rHA protein of the H5N1 influenza virus was performed in both an aqueous buffer and human serum with a linear range of 0–100 ng mL^{-1} (figure 5.4(c)) and with LODs of 2.0 and 3.5 ng mL^{-1}, respectively. The specificity of the biosensor was far superior when testing other serum proteins (figure 5.4(d)).

A plasmonic fluorophore (figure 5.5(a)) was constructed using a bovine serum albumin scaffold with approximately 210 IRDye 800CW: a near-infrared fluorescent dye from Li-Core (with a fluorescence intensity approximately 6700 times that of a

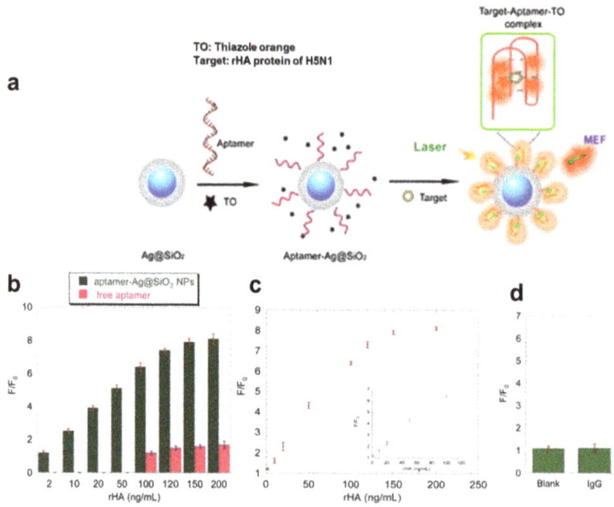

Figure 5.4. (a) Schematic illustration of the aptamer-Ag@SiO_2 sensor preparation and the analysis of the rHA protein of H_5N_1. (b) The F/F_0 of the aptamer alone in the solution (red) and the aptamer functionalized Ag@SiO_2 NPs with the addition of rHA protein (c) Fluorescence response of the aptamer-Ag@SiO_2 NPs solution (with 100 nM TO) upon addition of rHA (0–200 ng mL^{-1}). The figure inset in (c) shows the linear response with rHA concentrations from 0 to 100 ng mL^{-1}. (d) Selectivity test for rHA protein using rHA at a concentration of 100 ng mL^{-1}, IgG at a concentration of 100 μg mL^{-1}, and HSA at a concentration of 100 μ g mL^{-1}. Reprinted from [31], Copyright 2015, with permission from Elsevier.

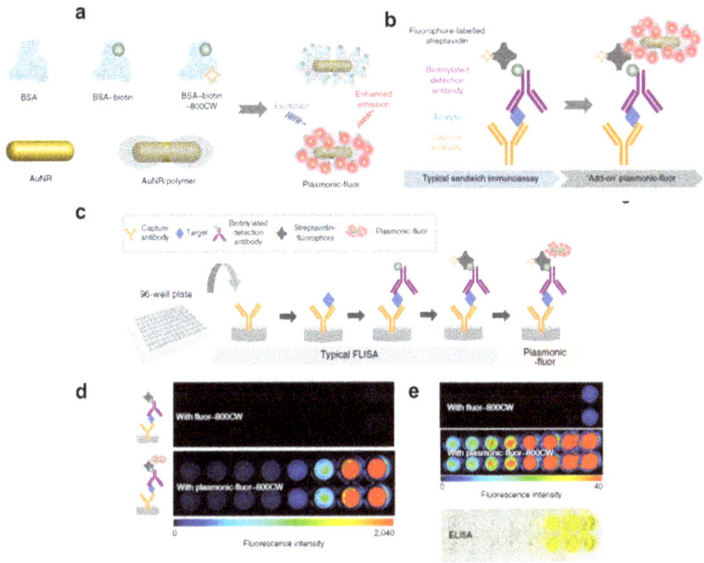

Figure 5.5. Overview of immunoassays with plasmonic fluor materials and p-fluorescence-linked immunosorbent assay (FLISA) and multiplexed bead. (a) Schematic illustration showing the structure of plasmonic fluor, which consists of a plasmonically active core (for example, an AuNR), a polymer shell as a spacer layer, a light emitter, and biotin. (b) The working principle of plasmonic fluor enhances the fluorescence intensity and consequent signal-to-noise ratio of fluorescence-based assays without changing existing assay workflows. (c) Schematic illustration showing the concepts of conventional FLISA and p-FLISA in a standard 96-well plate format. (d) Fluorescence intensity maps of human IL-6 FLISA and p-FLISA. (e) Fluorescent intensity maps (with magnified dose-dependent fluorescent intensity) from conventional FLISA. Reprinted by permission from Springer (GmbH): Springer Nature, Nature Biomedical Engineering [40], Copyright (2020).

single 800 CW fluorophore), a polymer-coated gold nanorod acting as a plasmonic antenna, and biotin as a biorecognition element. Its emission wavelength was tuned over the visible and near-infrared spectral regions by modifying its size, shape, and composition. This biosensor (figures 5.5(c) and (d)) improved the LOD in fluorescence-linked immunosorbent assays by up to 4750 times and is compatible with multiplexed bead-based immunoassays. This method offered faster assay times (up to 20 min) and lower sample volumes, as shown by the detection of a pro-inflammatory cytokine in mouse interstitial fluid and of urinary biomarkers in patient samples [40]. In plasmonic fluors, AuNPs absorb then scatter high concentrations of light, which causes fluorophores to emit much brighter signals. The high intensity of the signals means fewer proteins are needed to generate the signals, inspiring a method by which a highly sensitive COVID-19 antibody test could be designed in the foreseeable future.

5.3 Molecularly imprinted polymers: an overview

MIPs are artificial recognition materials, also called plastic antibodies, that have been tailored by template-assisted polymer synthesis with remarkable pre-programmed affinity and selectivity towards their targets, which are usually defined as templates [41, 42]. The synthesis procedure normally consists of three steps called pre-polymerization, polymerization, and template removal, as shown in figure 5.6 [43]. In the pre-polymerization step, the template interacts with functional monomers to form a complex stabilized by molecular interactions, and substantially 'frozen' by polymerization in the presence of functional and cross-linking monomers. Following polymerization, the template is removed from the three-dimensional polymeric matrix using simple solvent extraction, thereby creating binding sites capable of recognizing the template and its analogs specifically [42].

Due to their outstanding features, such as high selectivity and sensitivity, cost-effectiveness, robustness, reusability, and reproducibility, along with their temperature and pH stability, MIPs have broad application in many fields, such as separation/chromatography [44, 45], sensors [46, 47], drug delivery [48, 49], bioimaging/theranostics [49–51], and catalysis [52, 53].

5.3.1 Molecularly imprinted polymers for fluorescence-based virus detection

Viruses form a very interesting field of research within the wider scope of molecular imprinting due to their relatively small size and relevance to health care [54]. In addition to the aforementioned applications, MIPs offer promising potential for the recognition/detection of viruses [55–63]. However, they are mostly related to the other types of sensor (optical and electrochemical biosensors), rather than fluorescence biosensors. By comparison, fluorescence detection is not as straightforward as other detection methods, such as electrochemical, QCM, and surface plasmon

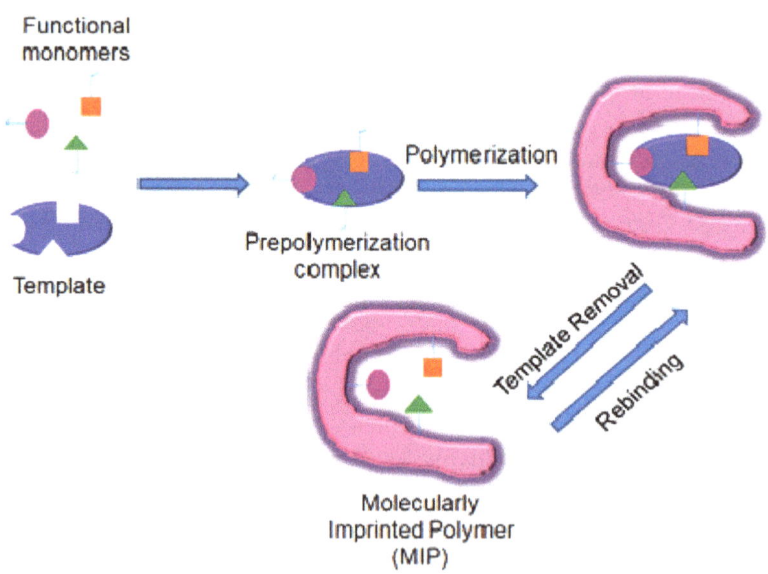

Figure 5.6. Schematic illustration of MIP synthesis. Reproduced from [43], with permission from MDPI. CC BY 4.0.

resonance. In particular, due to their relatively large dimensions, there are still challenging issues in the accurate quantification of viruses by fluorescent probes. A combination of MIP-based fluorescence detection and microfluidic flow cell cytometry could address these issues [64]. Cai's group has published several papers over five years addressing the use of MIP-based fluorescence sensors for the sensitive detection of viruses, especially Japanese encephalitis virus (JEV). In this regard, the first MIP-based fluorescent biosensor for virus detection was reported by Liang *et al* [65]. In this inspiring study, a MIP-based fluorescent strategy was proposed for the highly specific recognition and sensitive detection of JEV. Liang *et al* used JEV as the template for virus MIPs; a MIP layer was anchored to the surfaces of fluorescent silica microspheres which were previously modified using a fluorescent dye, pyrene-1-carboxaldehyde (PC). The presence of JEV was detected using FRET, which has been mostly reported as fluorescence quenching [66]. However, in this study, interestingly, the fluorescence intensity of PC was enhanced by the virus, as the virus and PC functioned as an energy donor and an energy acceptor, respectively [67]. The enhanced fluorescence intensity of the sensitive sensor was proportional to the concentration of JEV in the range of 24–960 pM, with an LOD of 9.6 pM.

In another study dealing with MIP-based sensitive fluorescent detection of JEV, Feng *et al* employed the surface molecular imprinting approach [68]. Initially, dansyl chloride (DNS-Cl), was covalently immobilized on silica microspheres for fluorescence signaling due to its excellent fluorescent properties. Following this immobilization, JEV-MIPs were prepared using JEV as the template, (3-aminopropyl) triethoxysilane as the monomer, and tetraethyl orthosilicate as the crosslinker. The resulting JEV-MIPs could selectively recognize JEV in the presence of the hepatitis A virus (HAV), simian virus 40, and rabies virus, and also sensitively detected JEV with a picomolar LOD in an hour.

A fluorescence MIP sensor based on a metal–organic framework (MOF) was shown to recognize JEV rapidly with high sensitivity and selectivity by Yang *et al* [69]. Zinc acrylate was used as a functional monomer to form MIPs on the surface of a silicon-modified MOF (MIL-101). Subsequently, inspired by the work of Mizaikoff's team [70], polyethylene glycol (PEG) was used as a passivation agent to enhance the selectivity and ability of polymers to specifically recognize the template JEV. The strategy used for the synthesis of MIPs and JEV detection is illustrated in figure 5.7.

The sensor exhibited a wide range of detection from 50 pM to 1400 pM, with an LOD of 13 pM in 20 min under optimum experimental conditions. The developed sensor was able to achieve high-sensitivity and -specificity detection of JEV in human serum.

Recently, Cai's group proposed a fluorescent sensor based on an MOF for the detection of weakly fluorescent HAV [71]. A luminescent MOF (MIL-101-NH_2) was utilized as the fluorescence output signal generator and imprinting carrier, which was formed onto the surfaces of magnetic Fe_3O_4 nanoparticles (figure 5.8). Subsequently, the template virus, e.g., HAV, was immobilized on the surface of the magnetic MOF, and an imprinting layer was formed on the surface of the Fe_3O_4 via hydrolysis using tetraethyl orthosilicate (TEOS) as the functional monomer. The

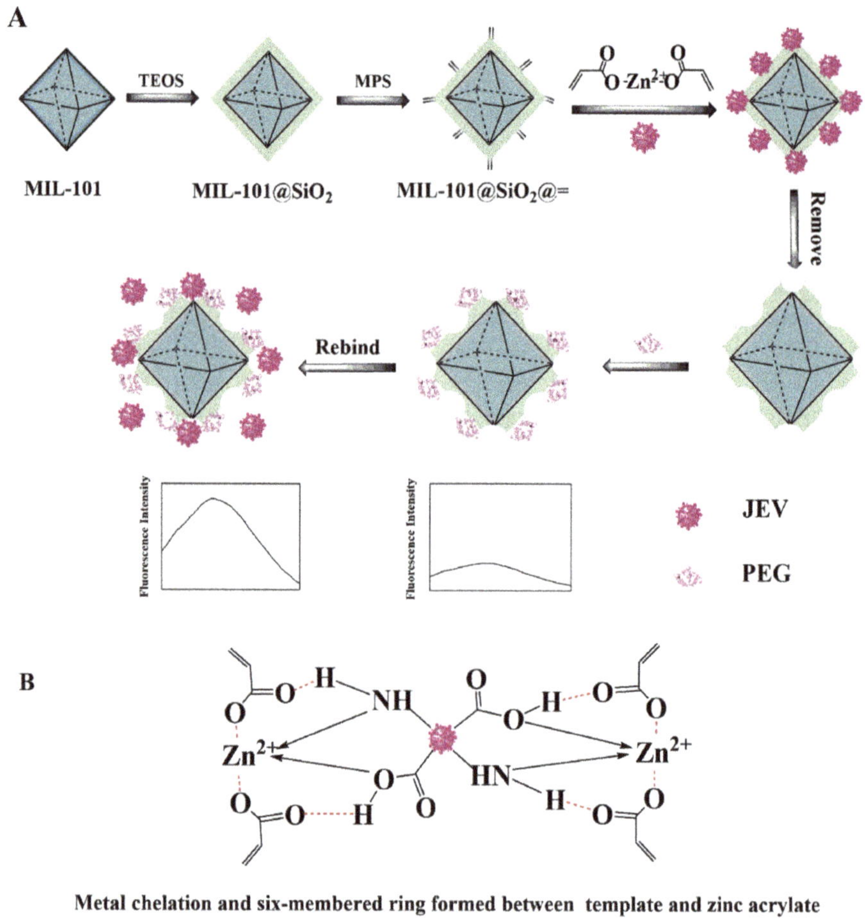

Figure 5.7. The strategy used for the synthesis of MIPs and JEV detection. Reprinted from [69], Copyright 2020, with permission from Elsevier.

resulting magnetic MIP integrated fluorescent sensor was used for the detection of HAV in human serum samples. The results revealed that the designed fluorescence sensor could be a promising sensor platform for quick, sensitive, and selective detection of HAV in the range of 0.02–2.5 nM with an LOD of 3.0 pM.

Cai's team recently developed a fluorescent MIP sensor based on a combination of aptamer and MOF for the sensitive and selective recognition of the hepatitis B virus (HBV) [72]. Viral aptamers were introduced onto the surface of MIL-101-NH_2 via an amide reaction following a surface imprinting approach carried out through the self-polymerization of TEOS. In this work, MIL-101-NH_2 was employed as the reference fluorescence signal ($\lambda_{ex}/\lambda_{em}$ = 290/460 nm) and rhodamine B as the change signal ($\lambda_{ex}/\lambda_{em}$ = 550/570 nm). As confirmed by the results of this study, the combination of ratiometric fluorescence detection with a dual recognition strategy not only increased the selectivity and the anti-interference ability of the sensor but

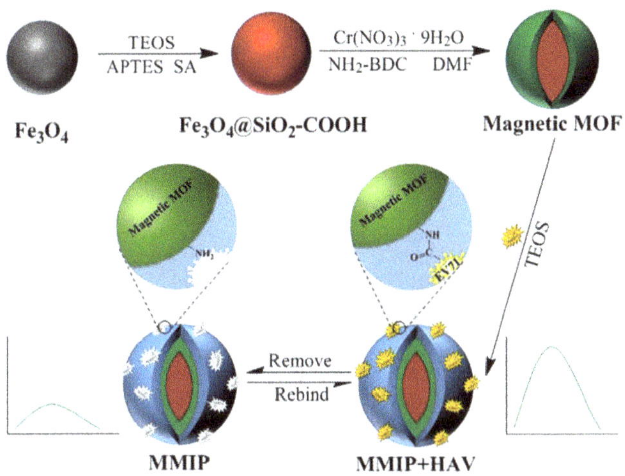

Figure 5.8. Schematic depiction of the synthesis of magnetic MIPs and HAV detection. Reprinted from [71], Copyright 2021, with permission from Elsevier.

also lowered the LOD to 1.8 pM, which could be a very promising approach for the specific detection of viruses in the future.

5.4 Conclusions and future perspectives

As the world has been struggling with rampant COVID-19 since 2020, it is now well understood that rapid, robust, highly sensitive, and selective biosensors are now, more than ever, of the utmost importance for the detection of viruses. In particular, the current health emergencies have highlighted an increasing need for novel biosensors that can match the selectivity and sensitivity of conventional nucleic acid tests. Recent years have seen tremendous efforts to develop optical biosensors for the detection of different viruses that are not only rapid, cheap, and easy to use but also sensitive and robust. This chapter has highlighted the current trends in fluorescence biosensors for virus detection in the light of recent literature. However, fluorescence biosensors, as a class of optical biosensors, still have a way to go for virus detection, compared to electrochemical and QCM biosensors.

In this context, nanomaterial/MIP-based fluorescence biosensors serve as ideal substitutes for conventional nucleic acid tests due to ongoing advancements in their design and fabrication methods, high selectivity, and sensitivity. There is no doubt that novel viral outbreaks will occur in the future, which is why it is crucial to harness fluorescence biosensing technology to aid ultrafast virus detection and prevent virus transmission.

References

[1] https://covid19.who.int

[2] Song M, Yang M and Hao J 2021 Pathogenic virus detection by optical nanobiosensors *Cell Rep. Phys. Sci.* **2** 100288

[3] La Scola B *et al* 2008 The virophage as a unique parasite of the giant mimivirus *Nature* **455** 100–4

[4] Saylan Y, Erdem Ö, Ünal S and Denizli A 2019 An alternative medical diagnosis method: biosensors for virus detection *Biosensors* **9** 65

[5] Abid S A, Muneer A A, Al-Kadmy I M S, Sattar A A, Beshbishy A M, Batiha G E and Hetta H F 2021 Biosensors as a future diagnostic approach for COVID-19 *Life Sci.* **273** 119117

[6] Afzal A, Mujahid A, Schirhagl R, Bajwa S Z, Latif U and Feroz S 2017 Gravimetric viral diagnostics: QCM based biosensors for early detection of viruses *Chemosensors* **5** 7

[7] Payungporn S, Chutinimitkul S, Chaisingh A, Damrongwantanapokin S, Buranathai C, Amonsin A, Theamboonlers A and Poovorawan Y 2006 Single step multiplex real-time RT-PCR for H5N1 influenza A virus detection *J. Virol. Methods* **131** 143–7

[8] Viswanathan R *et al* 2019 Comparison of two commercial ELISA kits for detection of rubella specific IgM in suspected congenital rubella syndrome cases and rubella IgG antibodies in a serosurvey of pregnant women *Diagn. Microbiol. Infect. Dis.* **94** 243–7

[9] Gourinat A, O'Connor O, Calvez E, Goarant C and Dupont-Rouzeyrol M 2015 Detection of Zika virus in urine *Emerg. Infect. Dis.* **21** 84–6

[10] Saylan Y, Yılmaz F, Özgür E, Derazshamshir A, Bereli N, Yavuz H and Denizli A 2018 Nanotechnology characterization tools for biosensing and medical diagnosis *Surface Plasmon Resonance Sensors for Medical Diagnosis* ed C S S R Kumar (Berlin: Springer) pp 425–58

[11] Zheng L, Cai G, Qi W, Wang S, Wang M and Lin J 2020 Optical biosensor for rapid detection of salmonella typhimurium based on porous gold@platinum nanocatalysts and a 3D fluidic chip *ACS Sens.* **5** 65–72

[12] Yang A and Yan F 2021 Flexible electrochemical biosensors for health monitoring *ACS Appl. Electron. Mater.* **3** 53–67

[13] Narita F, Wang Z, Kurita H, Li Z, Shi Y, Jia Y and Soutis C 2021 A review of piezoelectric and magnetostrictive biosensor materials for detection of COVID-19 and other viruses *Adv. Mater.* **33** 2005448

[14] Mauriz E 2020 Recent progress in plasmonic biosensing schemes for virus detection *Sensors* **20** 4745

[15] Qin P *et al* 2019 Rapid and fully microfluidic Ebola virus detection with CRISPR-Cas13a *ACS Sens.* **4** 1048–54

[16] Mertens J, Cuervo A and Carrascosa J L 2019 Nanomechanical detection of *Escherichia coli* infection by bacteriophage T7 using cantilever sensors *Nanoscale* **11** 17689–98

[17] Jin X, Zhang H, Li Y, Xiao M, Zhang Z, Pang D, Wong G, Zhang Z and Zhang G 2019 A field effect transistor modified with reduced graphene oxide for immunodetection of Ebola virus *Microchim. Acta* **186** 223

[18] Arranz A and Ripoll J 2015 Advances in optical imaging for pharmacological studies *Front. Pharmacol.* **6** 189

[19] Maddali H, Miles C E, Kohn J and O'Carroll D M 2021 Optical biosensors for virus detection: prospects for SARS-CoV-2/COVID-19 *ChemBioChem* **22** 1176–89

[20] Shrivastav A M, Cvelbar U and Abdulhalim I 2021 A comprehensive review on plasmonic-based biosensors used in viral diagnostics *Commun. Biol.* **4** 70

[21] Chan J, Dodani S C and Chang C J 2012 Reaction-based small-molecule fluorescent probes for chemoselective bioimaging *Nat. Chem.* **4** 973–84

[22] Cotruvo J A, Aron A T, Ramos-Torres K M and Chang C J 2015 Synthetic fluorescent probes for studying copper in biological systems *Chem. Soc. Rev.* **44** 4400–14
[23] Sharma A, Khan R, Catanante G, Sherazi T A, Bhand S, Hayat A and Marty J L 2018 Designed strategies for fluorescence-based biosensors for the detection of mycotoxins *Toxins* **10** 197
[24] Clegg R M 1995 Fluorescence resonance energy transfer *Curr. Opin. Biotechnol.* **6** 103–10
[25] Hong G, Robinson J T, Zhang Y, Diao S, Antaris A L, Wang Q and Dai H 2012 In vivo fluorescence imaging with Ag_2S quantum dots in the second near-infrared region *Angew. Chem. Int. Ed.* **51** 9818–21
[26] Fothergill S M, Joyce C and Xie F 2018 Metal enhanced fluorescence biosensing: from ultraviolet towards second near-infrared window *Nanoscale* **10** 20914–29
[27] Li J, Li C and Aroca R F 2017 Plasmon-enhanced fluorescence spectroscopy *Chem. Soc. Rev.* **46** 3962–79
[28] Inci F, Tokel O, Wang S, Gurkan U A, Tasoglu S, Kuritzkes D R and Demirci U 2013 Nanoplasmonic quantitative detection of intact viruses from unprocessed whole blood *ACS Nano* **7** 4733–45
[29] Takemura K, Adegoke O, Takahashi N, Kato T, Li T, Kitamoto N, Tanaka T, Suzuki T and Park E Y 2017 Versatility of a localized surface plasmon resonance-based gold nanoparticle-alloyed quantum dot nanobiosensor for immunofluorescence detection of viruses *Biosens. Bioelectron.* **89** 998–1005
[30] Hu J, Jiang Y, Wu L, Wu Z, Bi Y, Wong G, Qiu X, Chen J, Pang D and Zhang Z 2017 Dual-signal readout nanospheres for rapid point-of-care detection of Ebola virus glycoprotein *Anal. Chem.* **89** 13105–11
[31] Pang Y, Rong Z, Wang J, Xiao R and Wang S 2015 A fluorescent aptasensor for H5N1 influenza virus detection based-on the core–shell nanoparticles metal-enhanced fluorescence (MEF) *Biosens. Bioelectron.* **66** 527–32
[32] Iswardy E, Tsai T, Cheng I, Ho T, Perng G and Chang H 2017 A bead-based immunofluorescence-assay on a microfluidic dielectrophoresis platform for rapid dengue virus detection *Biosens. Bioelectron.* **95** 174–80
[33] Ming K, Kim J, Biondi M J, Syed A, Chen K, Lam A, Ostrowski M, Rebbapragada A, Feld J J and Chan W C W 2015 Integrated quantum dot barcode smartphone optical device for wireless multiplexed diagnosis of infected patients *ACS Nano* **9** 3060–74
[34] Xiong L, Cui R, Zhang Z, Yu X, Xie Z, Shi Y and Pang D 2014 Uniform fluorescent nanobioprobes for pathogen detection *ACS Nano* **8** 5116–24
[35] Zhang J, Tian J, He Y, Chen S, Jiang Y, Zhao Y and Zhao S 2013 Protein-binding aptamer assisted signal amplification for the detection of influenza A (H1N1) DNA sequences based on quantum dot fluorescence polarization analysis *Analyst* **138** 4722–7
[36] Cai H *et al* 2015 Optofluidic analysis system for amplification-free, direct detection of Ebola infection *Sci. Rep.* **5** 14494
[37] Ashiba H, Sugiyama Y, Wang X, Shirato H, Higo-Moriguchi K, Taniguchi K, Ohki Y and Fujimaki M 2017 Detection of norovirus virus-like particles using a surface plasmon resonance-assisted fluoroimmunosensor optimized for quantum dot fluorescent labels *Biosens. Bioelectron.* **93** 260–6
[38] Hsieh B, Chang Y, Ng M, Liu W, Lin C, Wu H and Chou C 2007 Localized surface plasmon coupled fluorescence fiber-optic biosensor with gold nanoparticles *Anal. Chem.* **79** 3487–93
[39] Weng X and Neethirajan S 2018 Immunosensor based on antibody-functionalized MoS_2 for rapid detection of avian coronavirus on cotton thread *IEEE Sens. J.* **18** 4358–63

[40] Luan J *et al* 2020 Ultrabright fluorescent nanoscale labels for the femtomolar detection of analytes with standard bioassays *Nat. Biomed. Eng.* **4** 518–30
[41] Piletsky S, Canfarotta F, Poma A, Bossi A M and Piletsky S 2020 Molecularly imprinted polymers for cell recognition *Trends Biotechnol.* **38** 368–87
[42] Haupt K, Medina Rangel P X and Tse Sum Bui B 2020 Molecularly imprinted polymers: antibody mimics for bioimaging and therapy *Chem. Rev.* **120** 9554–82
[43] Yarman A and Scheller F W 2020 How reliable is the electrochemical readout of MIP sensors? *Sensors* **20** 2677
[44] Shinde S, Incel A, Mansour M, Olsson G D, Nicholls I A, Esen C, Urraca J and Sellergren B 2020 Urea-based imprinted polymer hosts with switchable anion preference *J. Am. Chem. Soc.* **142** 11404–16
[45] Xing R, Wen Y, He H, Guo Z and Liu Z 2019 Recent progress in the combination of molecularly imprinted polymer-based affinity extraction and mass spectrometry for targeted proteomic analysis *TrAC-Trends Anal. Chem.* **110** 417–28
[46] Garcia Cruz A, Haq I, Cowen T, Di Masi S, Trivedi S, Alanazi K, Piletska E, Mujahid A and Piletsky S A 2020 Design and fabrication of a smart sensor using in silico epitope mapping and electro-responsive imprinted polymer nanoparticles for determination of insulin levels in human plasma *Biosens. Bielectron.* **169** 112536
[47] Ahmad O S, Bedwell T S, Esen C, Garcia-Cruz A and Piletsky S A 2019 Molecularly imprinted polymers in electrochemical and optical sensors *Trends Biotechnol.* **37** 294–309
[48] Sanadgol N and Wackerlig J 2020 Developments of smart drug-delivery systems based on magnetic molecularly imprinted polymers for targeted cancer therapy: a short review *Pharmaceutics* **12** 831
[49] Qin Y, Feng Y, Ma Y, He X, Li W and Zhang Y 2020 Tumor-sensitive biodegradable nanoparticles of molecularly imprinted polymer-stabilized fluorescent zeolitic imidazolate framework-8 for targeted imaging and drug delivery *ACS Appl. Mater. Interfaces* **12** 24585–98
[50] Shinde S, El-Schich Z, Malakpour A, Wan W, Dizeyi N, Mohammadi R, Rurack K, Wingren A G and Sellergren B 2015 Sialic acid-imprinted fluorescent core–shell particles for selective labeling of cell surface glycans *J. Am. Chem. Soc.* **137** 13908–12
[51] Medina Rangel P X, Moroni E, Merlier F, Gheber L A, Vago R, Tse Sum Bui B and Haupt K 2020 Chemical antibody mimics inhibit cadherin-mediated cell-cell adhesion: a promising strategy for cancer therapy *Angew. Chem. Int. Ed.* 59 2816–22
[52] Li X and Zhao Y 2019 Chiral gating for size- and shape-selective asymmetric catalysis *J. Am. Chem. Soc.* **141** 13749–52
[53] Muratsugu S, Shirai S and Tada M 2020 Recent progress in molecularly imprinted approach for catalysis *Tetrahedron Lett.* **61** 151603
[54] Hayden O, Lieberzeit P A, Blaas D and Dickert F L 2006 Artificial antibodies for bioanalyte detection—sensing viruses and proteins *Adv. Funct. Mater.* **16** 1269–78
[55] Yang B, Gong H, Chen C, Chen X and Cai C 2017 A virus resonance light scattering sensor based on mussel-inspired molecularly imprinted polymers for high sensitive and high selective detection of hepatitis A virus *Biosens. Bioelectron.* **87** 679–85
[56] Altintas Z, Pocock J, Thompson K and Tothill I E 2015 Comparative investigations for adenovirus recognition and quantification: plastic or natural antibodies? *Biosens. Bioelectron.* **74** 996–1004

[57] Klangprapan S, Choke-arpornchai B, Lieberzeit P A and Choowongkomon K 2020 Sensing the classical swine fever virus with molecularly imprinted polymer on quartz crystal microbalance *Heliyon* **6** e04137
[58] Luo L, Zhang F, Chen C and Chai C 2019 Visual simultaneous detection of Hepatitis A and B viruses based on a multifunctional molecularly imprinted fluorescence sensor *Anal. Chem.* **91** 15748–56
[59] Gast M, Wondany F, Raabe B, Michaelis J, Sobek H and Mizaikoff B 2020 Use of super-resolution optical microscopy to reveal direct virus binding at hybrid core–shell matrixes *Anal. Chem.* **92** 3050–7
[60] Gast M, Sobek H and Mizaikoff B 2019 Selective virus capture via hexon imprinting *Mater. Sci. Eng.* C **99** 1099–104
[61] Dinc M, Esen C and Mizaikoff B 2019 Recent advances on core–shell magnetic molecularly imprinted polymers for biomacromolecules *TrAC-Trend. Anal. Chem.* **114** 202–17
[62] Gast M, Sobek H and Mizaikoff B 2019 Advances in imprinting strategies for selective virus recognition a review *TrAC-Trend. Anal. Chem.* **114** 218–32
[63] Gast M, Kühner S, Sobek H and Mizaikoff B 2018 Understanding the viral load during the synthesis and after rebinding of virus imprinted particles via real-time quantitative PCR *Analyst* **143** 2616–22
[64] Dar K K, Shao S, Tan T and Lv Y 2020 Molecularly imprinted polymers for the selective recognition of microorganisms *Biotechnol. Adv.* **45** 107640
[65] Liang C, Wang H, He K, Chen C, Chen X, Gong H and Cai C 2016 A virus-MIPs fluorescent sensor based on FRET for highly sensitive detection of JEV *Talanta* **160** 360–6
[66] Richards C I, Hsiang J, Khalil A M, Hull N P and Dickson R M 2010 FRET-enabled optical modulation for high sensitivity fluorescence imaging *J. Am. Chem. Soc.* **132** 6318–23
[67] Borah R and Kumar A 2016 Fluorescence enhancement of glutaraldehyde functionalized polyaniline nanofibers in the presence of aromatic aminoacids *Mat. Sci. Eng.* C **61** 762–72
[68] Feng W, Liang C, Gong H and Cai C 2018 Sensitive detection of Japanese encephalitis virus by surface molecularly imprinted technique based on fluorescent method *New J. Chem.* **42** 3503–8
[69] Yang J, Feng W, Liang K, Chen C and Cai C 2020 A novel fluorescence molecularly imprinted sensor for Japanese encephalitis virus detection based on metal organic frameworks and passivation-enhanced selectivity *Talanta* **212** 120744
[70] Gast M, Kühner S, Sobek H, Walther P and Mizaikoff B 2018 Enhanced selectivity by passivation: molecular imprints for viruses with exceptional binding properties *Anal. Chem.* **90** 5576–85
[71] Wang L, Liang K, Feng W, Chen C, Gong H and Cai C 2021 Molecularly imprinted polymers based on magnetically fluorescent metal–organic frameworks for the selective detection of hepatitis A virus *Microchem. J.* **164** 106047
[72] Wang L, Yang J, Tang L, Luo L, Chen C, Gong H and Cai C 2021 Specific determination of HBV using a viral aptamer molecular imprinting polymer sensor based on ratiometric metal organic framework *Microchim. Acta* **188** 221

IOP Publishing

Biosensors for Virus Detection

Adil Denizli and Yeşeren Saylan

Chapter 6

Thermal biosensors for virus detection

Huma Shaikh and Shahnila Shah

Biological reactions evolve or absorb heat. Thermal biosensors are based on measuring this property of biological reactions. Previously, conventional calorimeters were used to take advantage of different useful features of calorimetry in bioanalysis. Due to the more or less exothermic nature of biological reactions, the general calorimetric detection principle appears to be insensitive to the optical properties of samples. At present, viruses are a primary threat to human wellbeing, causing a range of different health issues, including infections ranging from minor to fatal. Therefore, the requirement for adequate detection methods is of chief importance in diverse areas, including the environment, medicine, food safety, and even counter-terrorism. Conventional diagnostic techniques, such as ELISA and PCR, are usually less cost- and time-effective and offer inadequate sensitivity and specificity. However, thermistor-based calorimetric measurements are reliable, cost-effective, and easy. In this chapter, thermometric biosensing principles are highlighted in the introduction and recent trends in thermal biosensing are also discussed in detail. The capability of thermal biosensors to detect viruses efficiently is also evaluated. In addition, a brief note on future developments in the field of thermal biosensors is provided.

6.1 Introduction

After the first explanation of biological thermodynamics by Hans Krebs in his book *Energy Transformations in Living Matter*, biological reactions were believed to be endothermic or exothermic [1]. Thus, the absorption or removal of heat is an important property of biological reactions. In both types of thermodynamic process, the reaction medium experiences a temperature change, and the study of this temperature change is known as calorimetry. Bioanalysis based on calorimetry is well established; it offers many advantages over other analytical techniques. The detection principle of calorimetry is universal, as it is not sensitive to the sample's optical properties, and depends on its heat change alone. Generally, the reactions of enzymes produce large changes of enthalpy, ranging between 20 and 100 kJ mol^{-1}.

doi:10.1088/978-0-7503-3867-7ch6

© IOP Publishing Ltd 2021

Therefore, it is convenient to measure changes using calorimetry based on only one enzymic step, as opposed to measuring other changes, for example, changes in the concentration of colored reactions. The broader applications of calorimetry in bioanalysis are precluded by their high cost, slow response, and reduced suitability for swift routine analysis [2]. In the beginning, calorimetry was used to study the extent of catalytic reactions and the dynamics of the structures of biomolecules in solutions [3]. However, it soon found its way into biosensors via thermometric devices [4]. Thermometric devices were mainly utilized to monitor the temperature changes of fluids that circulate during reactions between immobilized enzyme molecules and their appropriate substrates. The broad definition of thermometry is that it is a technique for monitoring temperature, and the most basic device based on the phenomenon of thermometry is the thermometer. Thus, the sensitive thermistors used in thermometric devices are based on similar principles [5]. The enzyme thermistor (ET) is a popular type of thermistor [6]. The development of ETs helped to overcome the drawbacks of calorimetric devices, i.e. their high operational cost and comparatively lengthy experimental measurements, as ET was associated with flow injection analysis in combination with a heat-sensing element and an immobilized biocatalyst [7]. Numerous instruments and devices have been designed in the last four decades that combine the principles of calorimetry, flow injection analysis, enzyme catalysis, and immobilization on suitable matrices.

Bioanalysis based on immobilized enzymes has motivated many researchers to develop such calorimeters for routine analysis that are simple, robust, reliable, and cost-effective. Their target was to develop clinical devices that could detect the metabolites of glucose and urea. Among different methods, the most effective and straightforward practice was the use of thermal enzyme probes (TEP). In this approach, the enzyme directly attaches to the thermistor or temperature transducer of the TEP [4, 8]. However, these thermistors offered low sensitivity, as a major portion of the enzymic reaction heat was lost to the surroundings without being detected by the thermistor. Pennington [9] proposed a low-volume calorimeter in which a Peltier element was used to sense the temperature of enzyme immobilized aluminum foil placed on its surface. He primarily proposed the TEP concept for batch operations. Even systems based on small columns with immobilized enzymes offered better sensitivity in the detection of reaction heat. In this procedure, the liquid that passed through the column helped to transport the heat along a thermistor that was situated on the top of the column. The 'enzyme thermistor' and [4, 10], the 'immobilized enzyme flow-enthalpimetric analyzer' are examples of this design [11]. A similar concept was proposed by Schmidt *et al* [12]. Muramatsu *et al* [13] designed a thermal biosensor probe with total dimensions of only $1 \times 1 \times 0.3$ mm for glucose detection as an integrated-circuit calorimetric sensor. Subsequently, a small thermoelectric glucose sensor was also developed that employed a thin-film thermopile to measure the evolved heat. This device appeared more stable than calorimetric sensors when responding to external thermal effects and could be operated without environmental temperature control [14].

6.2 Principle of thermal sensors

The principle of thermometric measurement is straightforward, as it only deals with the measurement of heat absorbed or evolved through a biochemical reaction. The evolution or absorption of the total heat is directly related to the amount of product formed and the molar enthalpy of the reaction.

$$Q = -n_p(\Delta H) \tag{6.1}$$

$$Q = C_p(\Delta T) \tag{6.2}$$

A description of equation (6.1) is as follows: the total heat is represented by Q, moles of the product are represented by n_p, and the molar enthalpy change is expressed by ΔH, which is is dependent on the heat capacity C_p of the reaction medium [3]. The temperature (ΔT) and enthalpy (ΔH) change are directly proportional to each other, and inversely proportional to the heat capacity of the system.

$$\Delta T = -\Delta H n_p / C_p \tag{6.3}$$

Organic solvents offer greater sensitivity and enhanced detection limits than those of aqueous solvents, because they have a heat capacity two or three times lower than those of aqueous systems, which results in unaltered ΔH [15]. The measurement of the change in the heat of a biochemical reaction is based on the sum of all the changes in the enthalpy of the reaction mixture. Hence, the sensitivity of the system can be improved by altering the biochemical reaction; for instance, by co-immobilizing oxidase with catalase or by using high-protonation enthalpy buffer ions such as Tris, the sensitivity of a thermistor can be doubled [16].

6.3 Recent trends in the field of thermal sensors

Since the discovery of thermistors, thermal sensors have evolved into very modern and sophisticated biosensors. The universal detection principle of thermal biosensors gives them broad application, especially in health care, as they are simple and easy to operate. The use of thermistors provides operational stability to biosensor systems and hence enhances their performance. For instance, the development of miniature thermal biosensor devices has proven useful in bioprocess industries. ET is also finding applications in bioprocess monitoring, leading to robust systems with sterile sampling. Thermal flow analyzers have attracted researchers to improve bioprocesses and downstream analysis. In this section, we will discuss some recent and modern forms of thermal biosensors.

6.3.1 Y-cut quartz resonator

A recent study reported the use of bulk acoustic wave resonators for the development of thermal biosensors [17]. Quartz crystal cuts that have a resonance frequency with a high temperature coefficient and very low phase noise are extremely sensitive thermometers, with an exceptional temperature resolution of up to 10^{-6} °C [18].

This work proposed the recording of the resonance frequency of developed micro-machined quartz resonators through an impedance analyzer. The measurement of the imaginary and real components of the impedance was performed and recorded for 801 frequency points within the available bandwidth. The resonators showed a temperature sensitivity of 7.32 kHz K^{-1} at around room temperature [19]. To monitor the resonance frequency of each pixel, impedance vs. frequency scans were performed. Approximately 600 ms was required for each 801 point impedance scan. While this method allows the determination of the resonance frequency for a broader temperature bandwidth, it is inherently a slow procedure, and changes in temperature occurring at rates ›1 s are difficult to track via this method. Figure 6.1 shows the temperature evolution temporal profile of a urea–urease reaction recorded using a Y-cut quartz resonator.

6.3.2 MEMS thermal biosensor

Microelectromechanical systems (MEMS) technology permits the cost-effective fabrication of miniaturized devices. The low sample volume, low thermal mass, and enhanced thermal isolation of this biosensor offer a broader linear range with tremendous sensitivity. It offers low power consumption, reduced analysis time and the parallel measurement of multiple samples.

The detection of temperature using sensors, e.g. thermistors or thermopiles, is the basis of MEMS thermal biosensors. Thermistors depend on changes in the electric resistance with temperature. Such changes can be used to determine the absolute temperature but with low sensitivity.

These biosensors are specifically used for gas and solid samples. Whereas, the use of these biosensors for liquid samples is limited to the detection of uric acid, glucose, and insulin and studies of living systems including animal cells, human cells, tissues, and micro-organisms. Currently, research is proceeding in the area of multiple-analyte-sensing thermal biosensors using MEMS thermopiles.

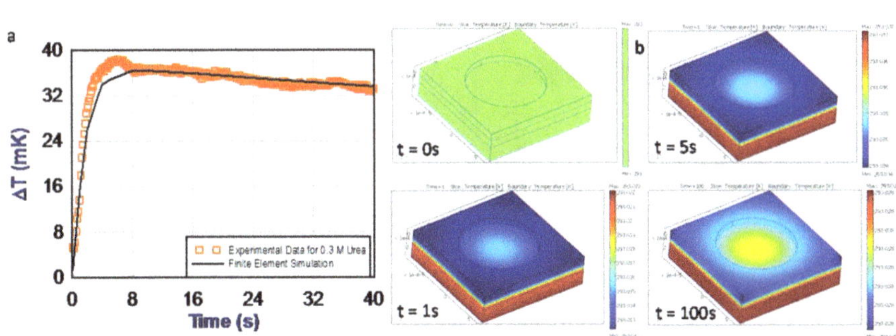

Figure 6.1. (a) Temporal profile of a 0.3 M urea–urease reaction. (b) The temperature profile's spatial distribution of pixels at different time intervals. Reproduced from [20] with permission of The Royal Society of Chemisty.

6.3.3 Single-walled carbon nanotube (SWCNT)-based soft thermal sensors

Yang *et al* [21] reported soft thermal sensors constructed by combining single-walled carbon nanotubes (SWCNTs) containing carboxyl groups with hydrogen-bond-based mechanically adaptable polymers. The plentiful H-bonds of the polymers confer excellent mechanical flexibility to soft thermal sensors and the ability to self repair repeatedly in a short time at room temperature (figure 6.2). Additionally,

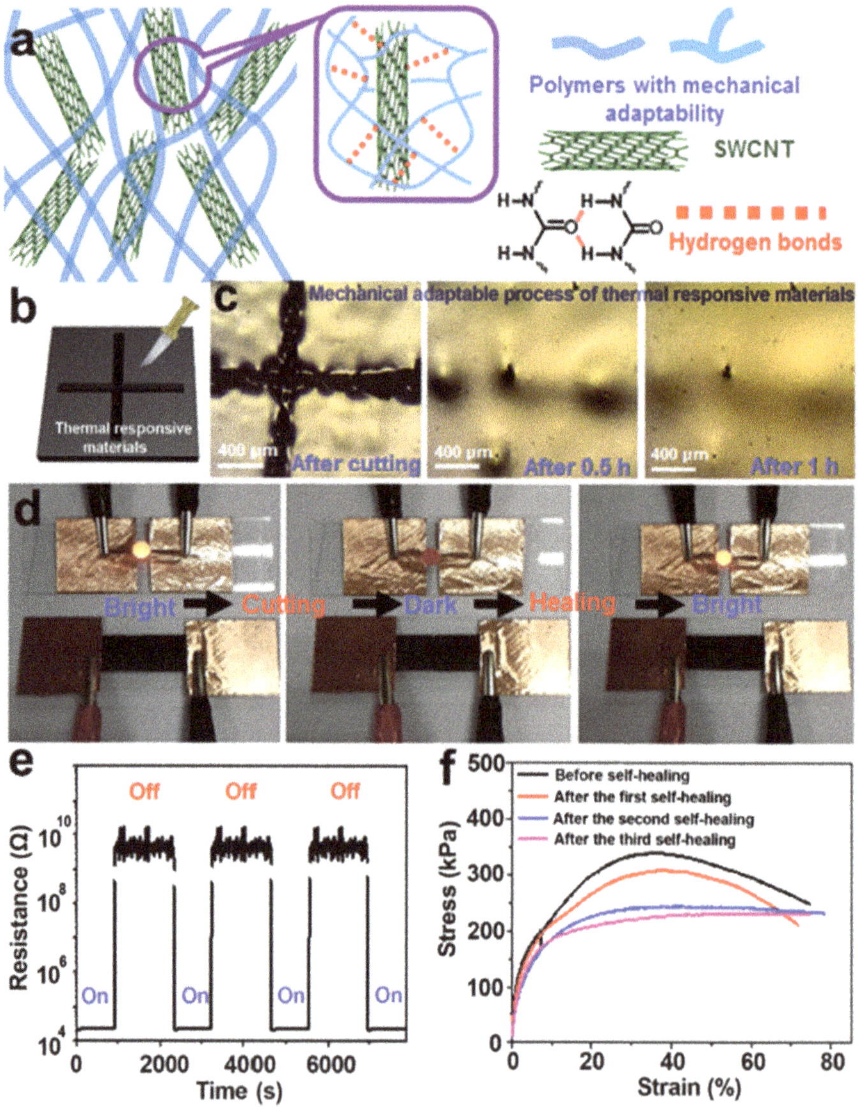

Figure 6.2. Thermally responsive materials: (a) schematic representation, (b) cross-shaped pattern carving, (c) microscopic photos, (d) electrical conductivity, (e) resistance recovery, and (f) tensile stress/strain curves. [21] John Wiley & Sons. Copyright © 1999-2021 John Wiley & Sons, Inc. All rights reserved.

covalent bonding between SWCNTs and polymers enhances the electrical conductivity of soft thermal sensors. Increasing the temperature of SWCNTs causes better contact, which decreases the thermal sensor resistance upon heating. Combining the electrical properties of SWCNTs and the thermal motion of polymer segments upon heating provides these soft materials with stable and repeatable thermal sensitivity. For this reason, incorporating non-covalent bonds with conductive SWCNTs endows soft thermal sensors with remarkable thermal sensitivity and outstanding mechanical adaptability, which will open up possibilities and challenges for soft materials, making them potential candidates for use in artificially intelligent robots and likely to replace the use of conventional hard materials in robots [22].

6.3.4 Thermoelectric biosensor (lab-on-a-chip)

Nestorova *et al* [23] proposed a hybrid lab-on-a-chip thermoelectric biosensor. A lab-on-a-chip is a miniaturized device that integrates routine laboratory analyses such as biochemical detection or DNA sequencing into a single chip. The inception of this miniaturized lab-on-a-chip field primarily depends on two main technologies: molecular biology and microfluidics. Microfluidic technologies allow the manufacture of millions of microchannels in a lab-on-chip device that help in the manipulation of biochemical reactions at very small volumes, as well as helping to facilitate the management of fluids in quantities as low as picoliters. In microfluidic calorimetry, the detection of biochemical reactions can be done by thermopiles owing to their suitability for miniaturization and their high common-mode thermal noise rejection ratio [24]. Figure 6.3 shows an antimony/bismuth thermopile. Thermopile-integrated microfluidic calorimeters open the door to several advantages, including a low sample volume requirement, label-free detection, high accuracy, and cost-effectiveness [25]. However, the output of a thermopile-integrated calorimeter relies on the reactant concentration

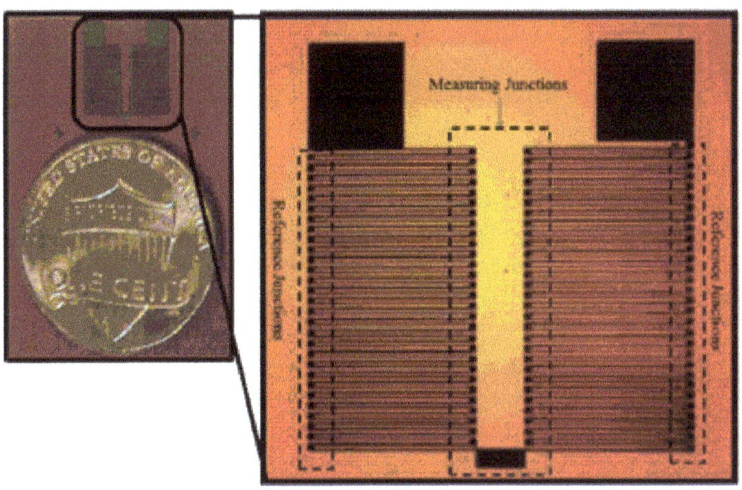

Figure 6.3. Thin-film antimony/bismuth thermopile. Reprinted from [23], Copyright 2016. with permission from Elsevier.

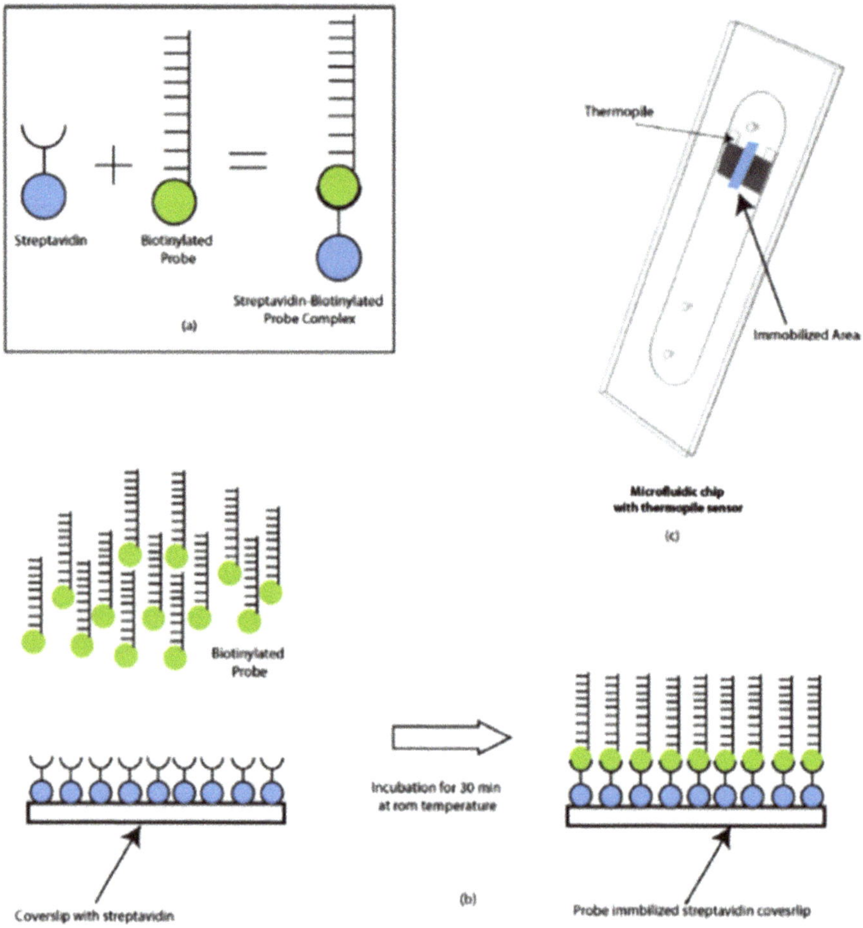

Figure 6.4. Immobilization of biotinylated nucleic acid probe on a streptavidin-coated glass coverslip and its working mechanism. Reprinted from [23], Copyright 2016. with permission from Elsevier.

and flow rate as well as the location of the reaction zone relative to the measuring junction of the thermopile [26]. Figure 6.4 schematically represents the sensing procedure of a lab-on-a-chip thermal biosensor.

Conventional bacterial investigative techniques take time and offer limited specificity and selectivity. To overcome this, highly sensitive molecular techniques have recently been introduced for selective and faster bacterial detection. Lately, more effective, sensitive molecular techniques have been developed by employing easier, cheaper, and stable synthetic receptors made by molecular imprinting techniques. Recently, the selective and specific label-free detection of *E. coli* bacteria using a biomimetic sensor was reported. This sensor measures the change in the thermal resistance surface when bacteria rebind to the polyurethane-coated aluminum chip-based synthetic cell receptor. The proposed thermal sensor can successfully detect bacterial rebinding and can also differentiate between dead and live bacteria [27].

6.4 Applications of thermal biosensors for virus detection

Thermal biosensors are probably one of the most accepted types of biosensor, due to their extensive ability to detect all variables in an embedded application. They can have a significant effect on materials and processes at the molecular level [28]. Owing to their stability and good resolution over periods of time, they have found numerous industrial and medical applications [29]. The potential of the thermal flow analyzer to be used in a multitude of applications is due to its universal detection principle. Thermal biosensors are currently attracting the special interest of researchers, due to their simple operating procedures and easy maintenance. Recent advancements have promoted an expansion in miniaturized thermal biosensor devices, which are useful in bioprocess industries due to their long-term stability, low cost, and high accuracy [30].

Over the last few decades, the detection of DNA through biosensing technology has grown rapidly. DNA biosensors can provide specific information regarding DNA sequencing quickly and simply and are widely applied for DNA-based detection and pathogen diagnosis [31]. The augmentation of exact genomic targets is required for the sensitive detection of nucleic acid sequences. While the fluorescent technique is generally used for the detection of amplified nucleic acid sequences, it is usually lacking in terms of sensitivity and requires labeling with different chemical reagents [32]. On the contrary, the label-free thermoelectric method measures the heat evolved during the hybridization reaction of DNA to provide direct information about the nucleic acid sequence. Gergana *et al* reported a novel thermoelectric DNA sensor for the detection of specific nucleic acid sequences for gene identification. Their study was based on the fabrication of a thermopile which was an integrated thin film of antimony–bismuth (Sb/Bi). It was efficient for the measurement of changes in temperature on the order of 10^{-4}K and offered high accuracy for complimentary and non-complimentary sequences [23]. As this thermosensor is appropriate for gene identification, it can be broadly applied for several applications. The thermosensor can be made specifically for the detection of a chosen virus.

Detecting viruses in the earlier stages of infection with high accuracy is of great significance in developing suitable medical treatments and preventing the further spread of infectious diseases. Diagnostic techniques based on conventional principles are usually very expensive and slow and offer inadequate sensitivity and specificity [33]. Recently, thermal imaging systems have been used to detect viruses through a virus detection application integrated into a wristband [34]. Infrared thermal cameras on the smart band (figure 6.5) perform temperature measurements. Consequently, the body's temperature at its surface is detected and captured by a thermal imaging camera.

6.5 Conclusions

As is evident from the literature, the field of thermal biosensors is growing swiftly day by day. Thermal biosensors have successfully been utilized for the detection of DNA. This development gives great hope that numerous thermal biosensors based on this phenomenon can be fabricated for the detection of a variety of viruses.

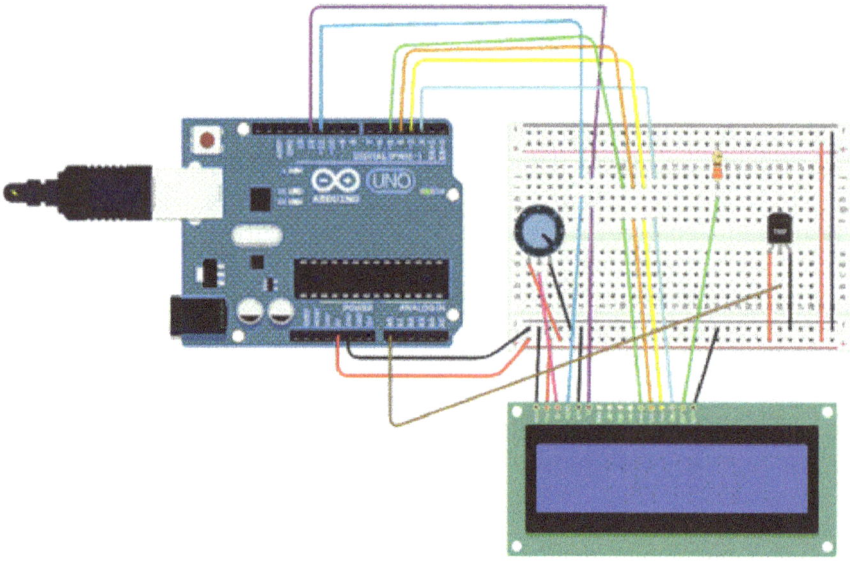

Figure 6.5. Temperature scanning module. Reprinted from [34], Copyright 2021. with permission from Elsevier.

Furthermore, soft thermal sensors can be turned into smart nanorobots that can detect viruses quickly. The combination of calorimetry with microfluidics has led to the development of lab-on-a-chip thermal biosensors. Lab-on-a-chip thermal biosensors can be fabricated for the detection of viruses by the non-technical user at the point of care, even at home, in the same way that a glucometer is used. However, until now, attention has not been given to this easy-to-use, specific, and swift sensing technology for the detection of viruses. Very few reports in the literature cover the detection of viruses using thermal biosensors. The technique of thermal biosensing is sufficiently developed, but there is a need to explore its wider use for virus sensing.

References

[1] Krebs H A and Kornberg H L 1957 *A Survey: Energy Transformations in Living Matter* (Berlin: Springer) p 87

[2] Spink C and Wadsö I 1976 Calorimetry as an analytical tool in biochemistry and biology *Methods of Biochemical Analysis.* pp 1–159

[3] Crime J K 1985 *Analytical Solution Calorimetry* (New York: Wiley)

[4] Mosbach K and Danielsson B 1974 An enzyme thermistor *Biochim. Biophys. Acta (BBA)—Enzymology.* **364** 140–45

[5] Danielsson B, Mattiasson B and Mosbach K 1981 Enzyme thermistor devices and their analytical applications *Applied Biochemistry and Bioengineering* ed L B Wingard, E Katchalski-katzir and L Goldstein (Amsterdam: Elsevier) pp 97–143

[6] Danielsson B and Mosbach K 1988 Enzyme thermistors *Methods in Enzymology* (New York: Academic) pp 181–97

[7] Turner A, Karube I and Wilson G S 1987 *Biosensors : Fundamentals and Applications.* 1st edn (Oxford: Oxford University Press) p 770

[8] Weaver J C *et al* 1976 Experiments and calculations concerning a thermal enzyme probe *Biochim. Biophys. Acta (BBA)—Enzymol.* **452** 285–91

[9] Pennington S N 1976 A small volume microcalorimeter for analytical determinations *Anal. Biochem.* **72** 230–37

[10] Danielsson B and Mosbach K 1988 Enzyme thermistors *Methods in Enzymology* (New York: Academic) pp 181–97

[11] Bowers L D and Carr P W 1976 An immobilized-enzyme flow-enthalpimetric analyzer: application to glucose determination by direct phosphorylation catalyzed by hexokinase *Clin. Chem.* **22** 1427–33

[12] Schmidt H L, Krisam G and Grenner G 1976 Microcalorimetric methods for substrate determination in flow systems with immobilized enzymes *Biochim. Biophys. Acta (BBA)—Enzymol.* **429** 283–90

[13] Muramatsu H, Dicks J M and Karube I 1987 Integrated-circuit bio-calorimetric sensor for glucose *Anal. Chim. Acta.* **197** 347–52

[14] Muehlbauer M J, Guilbeau E J and Towe B C 1989 Model for a thermoelectric enzyme glucose sensor *Anal. Chem.* **61** 77–83

[15] Ramanathan K, Jönsson B R and Danielsson B 2000 Analysis in Non-Aqueous Milieu Using Thermistors *Methods in Non-Aqueous Enzymology* ed M N Gupta (Basel: Birkhäuser Basel) pp 174–94

[16] Scheller F *et al* 1985 High-sensitivity enzyme thermistor determination of L-lactate by substrate recycling *Anal. Chem.* **57** 1740–3

[17] Lai S V H, Kao P and Tadigadapa S 2011 Thermal biosensors from micromachined bulk acoustic wave resonators *Procedia Eng.* **25** 1381–4

[18] EerNisse E P, Ward R W and Wiggins R B 1988 Survey of quartz bulk resonator sensor technologies *IEEE Trans. Ultrason. Ferroelectr. Freq. Control* **35** 323–30

[19] Kao P and Tadigadapa S 2009 Micromachined quartz resonator based infrared detector array *Sensors Actuators* A **149** 189–92

[20] Ren K *et al* 2011 Monitoring biochemical reactions using Y-cut quartz thermal sensors *Analyst* **136** 2904–11

[21] Yang H *et al* 2016 Soft thermal sensor with mechanical adaptability *Adv. Mater.* **28** 9175–81

[22] Cordier P *et al* 2008 Self-healing and thermoreversible rubber from supramolecular assembly *Nature* **451** 977–80

[23] Nestorova G G *et al* 2016 Lab-on-a-chip thermoelectric DNA biosensor for label-free detection of nucleic acid sequences *Sensors Actuators* B **225** 174–80

[24] Xie B *et al* 1994 Microbiosensor based on an integrated thermopile *Anal. Chim. Acta.* **299** 165–70

[25] Tangutooru S M *et al* 2012 Dynamic thermoelectric glucose sensing with layer-by-layer glucose oxidase immobilization *Sensors Actuators* B **166-7** 637–41

[26] Zhang Y and Tadigadapa S 2004 Calorimetric biosensors with integrated microfluidic channels *Biosens. Bioelectron.* **19** 1733–43

[27] van Grinsven B *et al* 2016 Label-free detection of escherichia coli based on thermal transport through surface imprinted polymers *ACS Sens.* **1** 1140–47

[28] Fontes J 2005 Temperature sensors *Sensor Technology Handbook* ed J S Wilson (Burlington: Newnes) pp 531–61

[29] Meijer G C M, Wang G and Heidary A 2018 Smart temperature sensors and temperature sensor systems *Smart Sensors and MEMs* 2nd edn ed S Nihtianov and A Luque (Cambridge: Woodhead Publishing) pp 57–85

[30] Vasuki S *et al* 2019 *Biosensors and Their Applications*

[31] Wang J 2000 Survey and Summary: From DNA biosensors to gene chips *Nucleic Acids Res.* **28** 3011–16

[32] Giller G *et al* 2003 Incorporation of reporter molecule-labeled nucleotides by DNA polymerases. I. Chemical synthesis of various reporter group-labeled 2'-deoxyribonucleoside-5'-triphosphates *Nucleic Acids Res.* **31** 2630–35

[33] Fournier P-E *et al* 2013 Modern clinical microbiology: new challenges and solutions *Nat. Rev. Microbiol.* **11** 574–85

[34] Jain S and Manocha A K 2021 Design and development of smart monitoring module for detection of virus *Measurement: Sensors* 100048

Chapter 7

Magnetic biosensors for virus detection

Tahira Qureshi, Muhammad Faisal, Sumera Qureshi and Najma Memon

Viruses have been observed to be a threat to human beings and other living beings, as they instigate harmful viral diseases. The current pandemic, which is due to severe acute respiratory syndrome-related coronavirus 2 (SARS-CoV-2) and which spreads by person-to-person contact, is also a viral disease, commonly known as COVID-19. Consequently, there is a notable need to identify viruses in a fast, selective, and precise manner. Previously, various pathogenic viruses were analyzed by commercial biosensors. Magnetic biosensors are developing in terms of their design and technical approach, and are proving an effective tool for virus detection. In this chapter, we discuss the current data and data from the last five years of research in the field of magnetic biosensors used for virus detection.

7.1 Introduction

Viruses are tiny organisms that are quite simple in structure compared to microbes. They require living cells to replicate, and are therefore intracellular parasites by nature. Viruses utilize the host cells' enzymes to synthesize modified viral species from their genes. Viruses are the root source of various infectious diseases, which can either be acute viral infections or chronic viral infections, such as ebola, influenza, chickenpox, flu, acquired immune deficiency syndrome (AIDS), severe acute respiratory syndrome (SARS), or the current pandemic coronavirus. The isolation or detection of viral antigens, deoxyribonucleic acid, or ribonucleic acid helps to confirm the presence of a specific virus in biological samples [1, 2]. The detection of these pathogenic viruses has an indispensable role in preventing the diseases they cause and developing effective immunizations against them. Up until now, viruses have been impossible to treat in comparison to other pathogens, for instance, bacteria or fungi, which can be treated with antibacterials and antifungals, respectively. In consequence, to combat viral infectious diseases in living beings, it is better to prevent a viral attack in the initial stages by the use of a specific immunization against the viral species. Furthermore, a precise mechanism is necessary to detect viruses and collect detailed

doi:10.1088/978-0-7503-3867-7ch7 © IOP Publishing Ltd 2021

information about viruses in order to fight viral diseases. Virus genetic materials consist of either deoxyribonucleic acid (DNA) or ribonucleic acid (RNA). The detection of these genetic materials has taken place using various methods, for instance, polymerase chain reactions (PCRs), reverse transcription-polymerase chain reactions (RT-PCRs), immunoblotting, viral flow cytometry, enzyme-linked immunosorbent assay (ELISA), etc. However, clinical sample collection, the isolation approach, and the inoculation of cell cultures of viruses have varying routes, as described in the literature. Also, these techniques for the identification or isolation of virus species are quite tedious, prolonged, and expensive processes [3–5]. To obtain a user-friendly and quick response, convenient and frugal techniques are needed for virus assessment. The current pandemic of COVID-19 has proved the need for these characteristics to prevent the chaotic spread of viral disease. Numerous immobilization techniques, immunosensing schemes, and their application in various virus detection approaches have been reported [6, 7]. Biosensing tools are quite inspiring analytical tools for the identification of biomolecules that bind an analyte with a bioreceptor in an integrated mechanism. Biosensors are proposed as analytical tools that are distinctively proficient at delivering a rapid response, either in quantitative or qualitative analysis, utilizing the biological identification of a targeted analyte. Biosensor-supported analysis seems a fitting approach to the detection of tiny viral species. It is a label-free procedure with no requirement for costly and laborious labeling steps; it is robust, and has an affinity with sophisticated micro- and nanofabrication technologies, such as those used for magnetic biosensors [8–10]. A typical block diagram of the biosensors used for virus detection is shown in figure 7.1. The virus interacts with a bioreceptor which acts as a sensing component; it forwards the information to a transducer and then to a signal reader or readout device.

The biosensor approach based on magnetic nanoparticles (MNPs) has established a prominent spot, as compared to other biosensing processes. For instance, magnetic biosensors offer cost-effective, biocompatible, deterministic, and environment-friendly virus assays. Furthermore, biological matter innately has negligible magnetic forces; therefore, external magnetic fields can be applied to achieve a response [11–13]. In the following section, we will describe the working of magnetic biosensors and an experimental approach to a biological assay for virus detection or isolation from samples. In this chapter, we evaluate the use of magnetic particles (MPs) (on the μm–nm scales) either with modified fabrication or directly employed in different biosensors to detect pathogenic viruses, as described by current research papers and those from the past five years.

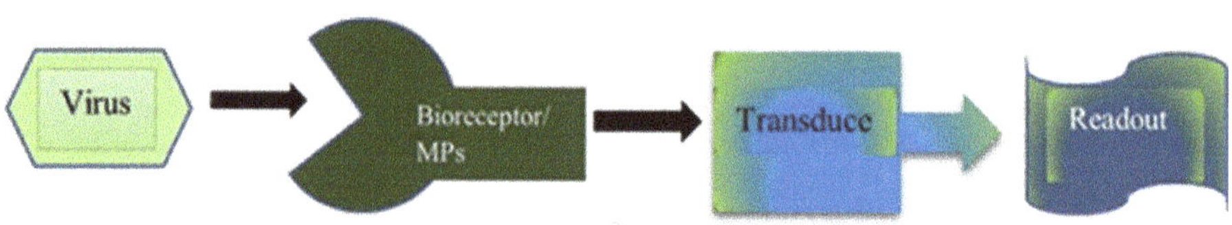

Figure 7.1. A block diagram of a biosensor.

7.2 Magnetic biosensors

Magnetic biosensors are utilized to indicate an alteration in a magnetically induced outcome or in their magnetic properties. A biological response is detected using superparamagnetic or paramagnetic crystals or particles by estimating changes in their magnetic characteristics by observing coil inductance or resistance. Fundamentally, magnetic biosensors can be categorized into substrate-based and substrate-free biosensors. Probes functionalized with MNPs bind directly to the biosensor's surface after the analyte is targeted, and such biosensors are called substrate-based biosensors. On the contrary, substrate-free biosensors utilize the resonance characteristic of MNPs by initiating a variation in the resonance behavior following the probe and target hybridization process. MP-based biosensors are unique in their approach to detecting the interaction of the analyte with MPs in terms of their signal-to-noise ratio, biosensor dimensions, the types of particle applied for detection, experimental approach, and amplification technique [13–15].

The distinctive characteristics of MNPs have facilitated effective assays of pathogens. MNPs have enabled secure target analysis by separating desired pathogenic species; this makes them efficient for ultrasensitive assays, as MPs can quantify samples from femtomolar to attomolar quantities. Biosensors using magnetism have established a prominent position due to their exceptional benefits as compared to other biosensing techniques. For instance, MNP preparation is cost-effective, their physical and chemical structures are stable, they offer biocompatibility for *in vitro* and *in vivo* studies, and MPs are also environment friendly. To date, various approaches have been applied and reported for the detection of viruses via magnetic biosensors. The scientific literature describes the use of magnetometers, magnetoresistive biosensors, Hall biosensors, and the direct use of MPs for virus detection. In addition, nuclear magnetic resonance (NMR) involves using magnetic nanoparticles for non-contact position detection by accelerating the relaxation rate of adjacent water molecules. This is a similar approach to magnetic resonance imaging (MRI) [16, 17].

Magnetic-particle-based biosensors are differentiated either by their signal output system or particle size, which leads to different characteristics depending on its dimensions (μm-nm). On the other hand, substrate-supported biosensors utilize uncovered MNPs which draw intense magnetic moments in response. MNPs labeled with specific biomolecular species have proven a suitable tool for various biorecognition assays [19, 20]. Figure 7.2 shows the measurement ranges of various magnetometers [18]. The next sections will introduce the different magnetic biosensor types and give an insight into the magnetic particles used to determine the presence of a virus species.

7.2.1 Magnetoresistive biosensors

Magnetoresistance (MR) is a change of resistance resulting from the application of a magnetic field. MR sensors utilize the dependence of the electric resistance on the magnetic field to perform analyses of conductors or semiconductors. However, improvements in the sensitivity of magnetic nanostructures have become possible due to the discoveries of the distinctive properties of antiferromagnetic exchange

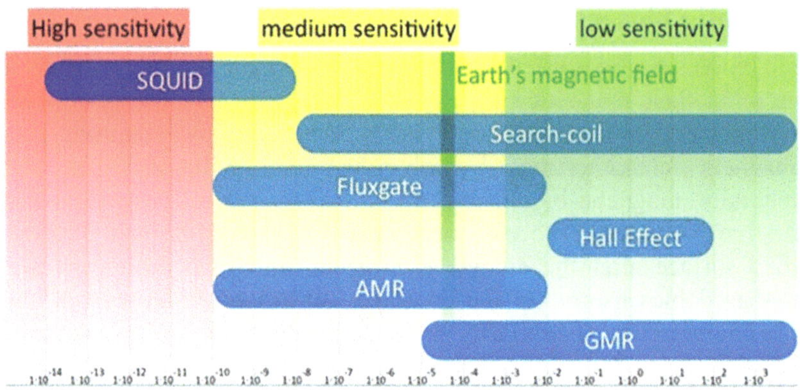

Figure 7.2. The measurement ranges of various magnetic sensors (in units of Tesla). Reprinted from [18] with author's permission.

coupling and giant magnetoresistance (GMR). In essence, MR dictates that the electrical resistance of a material is sensitive to the orientation of magnetization; examples include the GMR effect, the anisotropic MR (AMR) effect, and the tunneling MR (TMR) effect [21, 22].

7.2.2 Inductive biosensors

The inductive method senses the target molecule via the change of secondary coil inductance associated with the presence of superparamagnetic particles that appear close to the coil only when linked to the receptors. In general, this biosensor is constructed using a planar differential transformer with primary and secondary coils. The device's shape is symmetrical and divided into two parts; one part of the sensor is reactive and responsive to MNPs which capture the targeted biomolecule. The other part of the sensor acts as a dummy that is used to subtract spurious effects from the overall output of the transformer. MNPs fabricated in the coils or near the coils cause changes in the voltage induced in the biosensor. Here, the number of particles is proportional to the voltage induced by immobilizing the functionalized nanoparticles in the biosensor [23–25].

7.2.3 SQUID biosensors

Superconducting quantum interference devices (SQUIDs) are particularly helpful for recording low-frequency magnetic fluxes, such as biological currents originated by brain or heart movements. These biosensors operate with a quantum-limited sensitivity of 10^{-17} T at cryogenic temperatures. It is well documented that at cryogenic temperatures, some materials exhibit negligible electrical resistance, which makes them superconducting materials; on exposure to a magnetic field, such materials act as diamagnetic materials. Hence, when a superconducting material forms a circle, the magnetic field inside the circle is confined as a result of the cryogenic temperature. Such a circle contains a portion of non-superconducting material which reads out the tunneling current during the process.

These devices comprise a Josephson junction in a superconducting ring. The Josephson effect allows the measurement of minor changes in the inner magnetic field of an SQUID [26, 27].

7.2.4 Optomagnetic biosensors

Optomagnetic biosensors offer rapid, economical, sensitive, and quantitative analysis of the targeted biomolecule in solution, either by observing the light modulation of self-assembled MNPs, or by modifying their rotational behavior. The rotational dynamics of MNPs play a key role in optomagnetic biosensor detection. Brownian relaxation is a dominant factor in optomagnetic biosensors, as they use thermally insulated MPs. The optomagnetic biosensor approach based on magnetic particles can be used for the detection of a broad range of biological molecules and offers fast analysis. In this approach, MNPs are used to determine the alternating current (AC) resulting from magnetically excited field-induced modulation of the transformer; this provides fast and cost-effective analysis.

7.2.5 NMR biosensor

NMR biosensors can efficiently detect microbial pathogens with reproducibility and the ability to determine the presence of specific pathogenic species. Furthermore, the DNA of various micro-organisms has been detected in a variety of sample matrices with a rapid response time compared to those of classic approaches, which require a minimum of 24 h to produce results [28, 29].

NMR biosensors produce signals in milliseconds by determining the spin–spin relaxation time T_2; this depends on the size of the microorganism's DNA. Figure 7.3 shows the filter-based magnetic separation process, including the separation, washing, and back-flushing processes. A magnet is employed to achieve separation of antibody functionalized MNPs; in the washing step, the MNP-pathogen conjugates are retained; then, saturated analyte is eluted using a magnetic force and syringing [30].

7.2.6 Ferromagnetic biosensors

The combination of a volumetric sensing approach with homogeneous biosensor schemes is characteristically effective, straightforward, and has the advantage of relying on signal generation from the whole suspension volume [31]. Generally, as a result of their large surface area and the unique properties of nanoparticle labels, such sensors are an efficient choice for point-of-care diagnostic applications. The biosensor consists of a microwave circuit constructed from a slot line and a coplanar waveguide. The responsive part of the sensor consists of the slot line and waveguide, which can be functionalized with targeted analyte-attracting ligands. Hence, the recorded signal is inductively coupled to the responsive signal, with the result that the recorded signal is proportional to the response signal's frequency and intensified by the replicated ferromagnetic resonance (FMR) of the MNPs [32, 33].

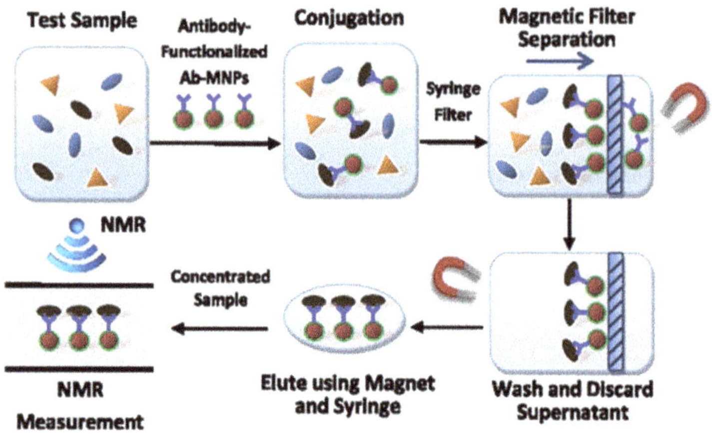

Figure 7.3. Working principles of NMR-based biosensors used for pathogen detection. Reprinted by permission from Springer (GmbH): Springer Nature, Journal of Biological Engineering [30], Copyright (2017).

7.2.7 Additional readout techniques

In addition to the methods described above, numerous other techniques have been used to detect pathogens using MPs. For instance, the fluxgate biosensor measures DC or low-frequency AC magnetic fields [34, 35]. Another related method is magnetic susceptibility, which uses a dimensionless proportionality constant that shows the degree of magnetization of a material in response to an applied magnetic field. This is caused by the interactions of electrons and nuclei with an externally applied magnetic field [36, 37]. Another related technique is the air-core magnetometer, which has applications in the biomedical field. In addition, nonlinear magnetic materials can also be detected by magnetic biosensors using different frequencies. In this method, two different frequencies of magnetic field are used to excite the material, which gives a linear and nonlinear frequency response that is a combination of the two separate excitation frequencies [38, 39].

7.3 Latest findings in the field of magnetic biosensors and virus detection

MPs have applications in a variety of fields; in particular, in biomedical-related biomolecular sensing and imaging, MPs are quite versatile in their uses, due to their wide range of roles, such as molecule-specific taggers, signal providers, or even force draggers. Consequently, their high magnetic strength, stability, biocompatibility, and flexibility with further tailored fabrication to support target analyte sensing make them a desirable approach for biosensing. The tailoring of MPs and surface enhancement with suitable functional groups are essential to achieve the desired molecule biosensing; however, exploration in this area is still ongoing. Early studies reported that size selection is one of the significant factors in the choice of MPs in different types of magnetometer process. It is well documented that larger MPs (>100 nm) are frequently employed in surface-based methods for the quantification of characteristics (for instance, in the GMR sensor and the force-induced remnant magnetization spectroscopy (FIRMS) technique), in which the larger particles play a role in enhancing signal-to-noise ratios. However, MPs smaller than 100 nm are extensively utilized in products based on the volumetric types of magnetometer, such as MRI imaging and SQUID-based immunoassays, in which smaller particle sizes are required to read low-frequency signals [40, 41]. In the following sections, developments in the magnetic biosensing of selected viruses over the past five years (2016–2020) are discussed, along with a few current insights.

7.3.1 Coronavirus

SARS is a highly infectious disease that results in the death of a fraction of patients [42]. The SARS disease is caused by the SARS coronavirus (SARS-CoV), which is evident in the respiratory secretions of patients after infection; it is commonly known as coronavirus. Initial-stage diagnosis and rapid sample screening are urgently required to control the geographical spread of the COVID-19 pandemic. Currently, two types of viral test are used: nucleic acid amplification tests (NAATs) and antigen tests. Serological testing gives information about previously encountered infections [42, 43]. Biosensor techniques offer a ray of hope that COVID-19 identification assays in clinical laboratories can be improved upon, since they offer speed, sensitivity, specificity, and a smaller number of manual steps; in addition, they take up less space.

7.3.2 Hepatitis

Hepatitis is caused by hepatotropic viruses and result in inflammation of the liver; it encompasses a broad range of acute and chronic liver diseases with various causes. Viral hepatitis belongs to a group of diseases comprising hepatitis viruses A (HAV), B (HBV), C (HCV), D (HDV), and E (HEV). The development of viral hepatitis into acute viral hepatitis has been observed for all hepatitis types [44].

Choi *et al* proposed a portable quantitative immunoassay approach utilizing a GMR biosensor. In their study, they performed whole-blood simultaneous detection of both human IgG and IgM antibodies; in addition, they employed smartphone integration. They claimed that their findings can be beneficial in the detection of IgG and IgM antibodies to specific hepatitis virus antigens, and can specify an infection status (acute or chronic) [45]. The combination of various other approaches with the use of magnetoresistive biosensors and magnetic markers has been proved to be an efficient approach for quantitative targeted molecule assay. For example, Sharma *et al* initially performed HEV, *Listeria monocytogenes*, and *Salmonella typhimurium* bacterium DNA extraction followed by detection. They first used a fluorescence technique and subsequently employed a magnetic platform for validation [46].

Altay *et al* developed an electrochemical evaluation using guanine oxidation in the presence of label-free HBV DNA hybridization; a disposable sensor approach was applied to the use of customized amino-functional carbon-coated magnetic nanoparticles (NH_2-CC-MNPs). Linear detection was achieved for DNA concentrations in the range of 5–25 μg mL^{-1} and the detection limit obtained was 1.15 μg mL^{-1} (20 pmol in 110 μl solution) [47]. Micheli *et al* proposed the detection of HAV in food samples via enzyme-linked immuno-magnetic electrochemical (ELIME) assay. They reported quantitative detection of HAV with a linear range of 10^{-11}–5×10^{-8} IU mL^{-1} and the detection limit achieved for the quantitative determination of HAV was $1 \cdot 10^{-12}$ IU mL^{-1} [48]. Alizadeh *et al* developed an electrochemical immunosensor that utilized a Au sheet as the working electrode, MNPs as a supporting matrix, and hemin/G-quadruplex DNA enzyme as a signal amplifier for the detection of hepatitis B virus surface antigen (HBsAg). The working range of the method developed for HBsAg recognition was 0.3–1000 pg mL^{-1} and the limit of detection (LOD) was 0.19 pg mL^{-1} [49]. Mashhadizadeh and Talemi reported a modified carbon paste electrode that used magnetite and gold nanoparticles for the immobilization of thiol-modified hepatitis B virus (HBV) probe DNA for the detection of trace amounts of HBV DNA. The DNA biosensor for HBV had an LOD of 3.1 (±0.1) $\times 10^{-13}$ M [50]. Huang *et al* developed a fluorescence-based aptasensor for hepatitis B e antigen (HBeAg); they modified HBeAg with magnetic beads to generate a target-specific response. The magnetic beads behaved as a homogeneous surface with different functional groups for the immobilization of the chosen analyte. The LOD obtained for HBeAg was 609 ng mL^{-1} [51]. Ghafary *et al* detected HBsAg by a surface-enhanced fluorescence (SEF) nanohybrid approach using covalent conjugate thionine–gold nanoparticles (GNP–Th) as an amplified fluorescence label, and MNPs as a biological transporter. The observed linear range was 4.6×10^{-9} to 0.012 ng mL^{-1} with an LOD of 4.6×10^{-9} ng mL^{-1} [52]. Yang *et al* utilized MNPs as a label in an immunoassay process; they reported a potassium ferrocyanide reaction with MNPs, resulting in *in situ* generation of Prussian blue. Furthermore, the generated Prussian blue showed catalytic activity in a luminol chemiluminescent (CL) reaction. They applied the developed method for the detection of carcinoembryonic antigen (CEA) and HBV-related sequence-specific DNA. The LODs for the detection of CEA and sequence-specific DNA were found to be 0.28 ng mL^{-1} and 0.044 pmol, respectively [53].

7.3.3 Human immunodeficiency virus

Human immunodeficiency virus (HIV) is a virus that attacks the body's immune system and causes chronic immune activation and oxidative stress. HIV remains a major global public health issue, having taken almost 33 million lives so far. With growing knowledge that AIDS causes a worldwide public health risk, various HIV assessment kits have been developed [54]. Low *et al* investigated the viral load and LILRA3 impact in the context of HIV infection. LILRA3 is an immunostimulatory molecule that also temporarily induces the propagation of cytotoxic cells. In this study, they purified a cluster of differentiation 14 (CD14) monocytes using commercially available magnetic particles (Becton–Dickinson) for magnetic-activated cell sorting. They proposed the possibility that LILRA3 could be manipulated for the treatment of viral diseases, such as HIV [55]. In another study, Phan *et al* used a temperature-sensitive MNP reagent system to pre-concentrate the HIV from samples using antiviral lectin cyanovirin-N from sample volumes of ~ 1 ml. They obtained approximately 100% of the HIV; around 1×10^5 viral copies were obtained from a sample of 100 µl in 30 min [56]. In another report, Damhorst *et al* used magnetic microparticles functionalized with antibodies or proteins targeted to capture HIV-1 virions from whole blood, capturing as many as 42.7% of the viruses in samples with high viremia [57]. Knowlton *et al* designed a multiplex platform for density-based cell sorting via magnetic focusing, which was combined with a smartphone for fluorescence imaging. The proposed device is compact and contains a smartphone and plastic components; the design is portable and easy to carry. They combined two techniques, magnetic focusing and fluorescence microscopy, and, as a result, created a point-of-care tool. They proposed a method for the identification and characterization of cell type and cell activity [58].

7.3.4 Ebola

Ebola virus (EBOV), a member of the *Filoviridae* family, is a negative sense ssRNA virus that has a filamentous form. Ebola virus disease is a serious, frequently fatal illness, with a death rate of up to 90%. Ebola affects humans and nonhuman primates, such as monkeys, gorillas, and chimpanzees [59]. Carinelli *et al* investigated the ebola virus (EBOV) using isothermal amplification of magnetic particles with an electrochemical genosensing technique. They studied ebola padlock probes to detect a specific *L*-gene sequence present in the five most common ebola species. Ebola complementary DNA (cDNA) was amplified by rolling circle amplification (RCA) on magnetic particles. The linear range of the electrochemical genosensor was 200 ymol to 200 zmol, corresponding to 120 cDNA molecules of *L*-gene; the LOD obtained was 33 cDNA molecules for EBOV detection [60]. Rosenstierne *et al* investigated EBOV on a miniaturized scale by performing multiple pathogen determination on a single chip. Another approach included magnetic-particle-based biosensors [61]. Zhao *et al* proposed a lateral flow biosensor (LFB) based on magnetic beads that performed enzyme-assisted isothermal strand-displacement amplification (SDA) to investigate four types of hemorrhagic fever virus: hantavirus (HTNV), chikungunya fever virus (CHIKV), dengue virus (DENV), and EBOV. The LOD of the DNA-dependent LFB was 10 fM [62].

7.3.5 Zika

Zika virus (ZIKV) is an arthropod-borne virus (arbovirus) of the family *Flaviviridae* and the genus *Flavivirus*. Zika virus disease is caused by a virus transmitted primarily by *Aedes* mosquitoes, which bite during the day [63]. Tian *et al* reported a synthetic ZIKV identification based on a combination of loop-mediated isothermal amplification (LAMP) with alternative current (AC) susceptometry analysis. The linear range obtained was 1 aM to 10 fM, and a 1 aM LOD of ZIKV oligonucleotide was observed within 27 min [64]. Danielli and his team developed magnetic modulation biosensing (MMB) for ZIKV detection. The sensitivity obtained for MMB was 88%–97%, while the specificity was 100%, and the cross-reactivities with West Nile and dengue viruses were also observed to be 0%–4%. Moreover, the MMB was able to analyze ZIKV IgM antibodies after just after five days and within a window of up to 180 days [65]. In a further study, Danielli *et al* investigated ZIKV by magnetic bead manipulation and mutual magnetization in a two-electromagnet setup process. They used electromagnets in a small footprint magnetic modulation biosensing system and obtained an LOD of 13 ng L^{-1} for recombinant ZIKV antibodies. This facilitated the determination of ZIKV IgM antibodies from as early as five days and up to 180 days, the same time window as in the previous study [66].

7.3.6 Norovirus

Noroviruses (NoVs) are members of the *Caliciviridae* family. There are five genera in this family, one of which is the NoV genus. NoV is a major cause of nonbacterial gastroenteritis outbreaks, and large numbers of illnesses happen each year [67]. Lee *et al* prepared binary nanoparticles, Au/MNP, which were enriched on graphenes (GRPs). Initially, for electrical sensing, the Au/MNP-GRPs were magnetically deposited on a Pt electrode; subsequently, the surface of the Au/MNP-GRPs was conjugated with norovirus antibody to generate a particle resembling NoV for specific sensing. The dynamic range for NoV detection was from 0.01 pg to 1 ng. The LOD was observed to be 1.16 pg mL^{-1} [68]. In another study, particles mimicking NoV were detected. This study used sandwiched approaches such as light signals detected from magnetic particles and optical markers [69].

7.3.7 Influenza virus

Influenza virus is a member of the *Orthomyxoviridae* family and contains ribonucleic acid (RNA) as its nucleic acid. The influenza A and B viruses are the main viruses that are highly contagious to humans and cause acute respiratory disease. RNA viruses have a spontaneous mutation rate compared to deoxyribose nucleic acid (DNA) viruses due to their lack of DNA polymerase I, which corrects errors that occur during DNA replication. Influenza viruses are transmitted by the inhalation of small-particle aerosols produced by coughing or sneezing [70]. Considering the characteristics of selective binding affinity between viral pathogens and cells, Chen *et al* proposed the detection of influenza viruses by wrapping the erythrocyte membrane of a selected pathogen cell in MNPs. The quantitative analysis was performed by qRT-PCR, an immunochromatography strip test, and cell-based titering analysis [71]. Jeun *et al* prepared magnetically engineered nanoferrite particles (MnMg-NFPs) and employed them in NMR to detect the influenza virus. The strength of the synthesized particles was regulated by the Mn and Mg concentrations. This modification of the Mn and Mg cations directly contributed to improving the transverse relaxation (R_2) rate [72].

7.4 Conclusions and future perspectives

MNP-based biosensors are frequently employed in biomedical applications and have become regular qualitative and quantitative tools for virus detection. Due to the use of biosensing coupled with microscopy and other detection approaches, MPs have started to provide rapid, convenient, selective, and sensitive detection/isolation of virus cells or their genetic materials (RNA/DNA). The main focus for future studies will be the optimization of MNPs in terms of surface chemistry, engineered shell mechanisms, and a controlled responsive dynamic range to regulate the magnetic response. It is well documented that MNPs can perform *in vivo* detection/isolation of viral cells or antigens for diagnosis. The detection of viruses is crucial for clinical and environmental applications. Additionally, samples containing low levels of viruses are a huge challenge for detection. Magnetic separation/isolation offers a substantial advantage for assessment by concentrating the sample manifold into small sample volumes and reducing interference in assays. The approach most used to enrich viral species from samples involves antibodies against the target virus on the surface of MNPs. To this end, antibody-conjugated MNPs have been used to separate human immunodeficiency virus (HIV-1) [73], *Salmonella typhimurium* [74], *Vibrio cholera*, and type 2 dengue virus. Peptides could be appropriate replacements for antibodies in the affinity-based separation of pathogens, now that, as a result of phage display technology, highly selective and specific peptides can be identified for target pathogens. In this manner, multiple antigenic peptides can be covalently immobilized on MNPs to capture herpes simplex virus 1, resulting in the sensitive and fast detection of this virus. The major purpose of advanced diagnostics is to deliver real-time detection of viruses using simpler, cost-effective, and sensitive methods. The advancement of nanotechnology has made revolutionary modifications in clinical diagnosis.

MNPs are showing encouraging outcomes in nanodiagnostics, although there is still room to find new varieties of MNP and ligands with which to perform real-time assays of biological samples. COVID-19 has brought attention to the development of real-time analysis with accurate outcomes from complex sample matrices. These fast techniques will allow physicians to instantly recommend a prescription to patients, decreasing the outcome time. Consequently, this will decrease patient load, improve conformity, and make the entire process cost-efficient. It has been reported that the magnetic bead (MB)-supported approach has played a significant role in the acquisition of targeted results for the isolation or extraction of specific biomolecules. For example, MBs coated with antibodies provide binding for specific lymphocyte cell surfaces; after binding with the targeted cell part in a magnetic field, control of the particle behavior is achieved. Recently, the development of MNP technology for the identification of viruses has been booming. The use of these particles in biosensors will probably enhance the mechanical, optical, and magnetic efficiencies following target analysis, and play a key role in increasing the precision and selectivity of virus recognition.

References

[1] Woolhouse M, Scott F, Hudson Z, Howey R and Chase-Topping M 2012 Human viruses: discovery and emergence *Philos. Trans. R. Soc. London* B **367** 2864–71

[2] Khan M A and Bass D M 2010 Viral infections: new and emerging *Curr. Opin. Gastroenterol.* **26** 26–30

[3] Saylan Y, Erdem Ö, Ünal S and Denizli A 2019 An alternative medical diagnosis method: biosensors for virus detection *Biosensors.* **9** 65

[4] Saleem M, Bilal M, Anwar S, Rehman A and Ahmed M 2013 Optical diagnosis of dengue virus infection in human blood serum using Raman spectroscopy *Laser Phys. Lett.* **10** 035602

[5] Samojlović M *et al* 2019 Detection of antibodies against lumpy skin disease virus by virus neutralization test and ELISA methods *Acta Vet.* **69** 47–60

[6] Tong A *et al* 2021 Research priorities for COVID-19 sensor technology *Nat. Biotechnol.* **39** 144–7

[7] Erdem Ö, Derin E, Sagdic K, Yilmaz E G and Inci F 2021 Smart materials-integrated sensor technologies for COVID-19 diagnosis *Emerg. Mater.* **4** 169–85

[8] Daaboul G G, Lopez C A, Yurt A, Goldberg B B, Connor J H and Ünlü M S 2011 Label-free optical biosensors for virus detection and characterization *IEEE J. Sel. Top. Quantum Electron.* **18** 1422–33

[9] Ozer T, Geiss B J and Henry C S 2019 chemical and biological sensors for viral detection *J. Electrochem. Soc.* **167** 037523

[10] Nguyen T H, Ung T D T, Vu T H, Tran T K C, Dinh D K and Nguyen Q L 2012 Fluorescence biosensor based on CdTe quantum dots for specific detection of H5N1 avian influenza virus *Adv. Nat. Sci.: Nanosci. Nanotechnol.* **3** 035014

[11] Rocha-Santos T A 2014 Sensors and biosensors based on magnetic nanoparticles *TrAC, Trends Anal. Chem.* **62** 28–36

[12] Rife J, Miller M, Sheehan P, Tamanaha C, Tondra M and Whitman L 2003 Design and performance of GMR sensors for the detection of magnetic microbeads in biosensors *Sens. Actuators,* A **107** 209–18

[13] Mokhtarzadeh A *et al* 2017 Nanomaterial-based biosensors for detection of pathogenic virus *TrAC, Trends Anal. Chem.* **97** 445–57

[14] Kim H H, Jeon H J, Cho H K, Cheong J H, Moon H S and Go J S 2015 Highly sensitive microcantilever biosensors with enhanced sensitivity for detection of human papilloma virus infection *Sens. Actuators* B **221** 1372–83

[15] Wang R *et al* 2011 Evaluation study of a portable impedance biosensor for detection of avian influenza virus *J. Virol. Methods* **178** 52–8

[16] Wang S X and Li G 2008 Advances in giant magnetoresistance biosensors with magnetic nanoparticle tags: review and outlook *IEEE Trans. Magn.* **44** 1687–702

[17] Xu Y and Wang E 2012 Electrochemical biosensors based on magnetic micro/nano particles *Electrochim. Acta* **84** 62–73

[18] Ettelt D 2012 *Conception and Fabrication of a Strain-Gauge Magnetometer* University of Grenoble

[19] Janssen X, Van Ijzendoorn L and Prins M 2008 On-chip manipulation and detection of magnetic particles for functional biosensors *Biosens. Bioelectron.* **23** 833–8

[20] Jamshaid T *et al* 2016 Magnetic particles: from preparation to lab-on-a-chip, biosensors, microsystems and microfluidics applications *TrAC, Trends Anal. Chem.* **79** 344–62

[21] Su D, Wu K, Saha R, Peng C and Wang J-P 2020 Advances in magnetoresistive biosensors *Micromachines.* **11** 34

[22] Lee J-R *et al* 2016 Experimental and theoretical investigation of the precise transduction mechanism in giant magnetoresistive biosensors *Sci. Rep.* **6** 1–9

[23] Beninato A *et al* 2016 Inductive integrated biosensor with extended operative range for detection of magnetic beads for magnetic immunoassay *IEEE Trans. Instrum. Meas.* **66** 348–59

[24] Andò B, Baglio S, Beninato A, Marletta V and Fallica G 2010 Development integrated inductive sensors for magnetic immunoassay in 'lab on chip' devices *From Physics to Control Through an Emergent View* (New York: World Scientific) pp 163–8
[25] Chai Y, Horikawa S, Wikle H C, Wang Z and Chin B A 2013 Surface-scanning coil detectors for magnetoelastic biosensors: a comparison of planar-spiral and solenoid coils *Appl. Phys. Lett.* **103** 173510
[26] Jianrong C, Yuqing M, Nongyue H, Xiaohua W and Sijiao L 2004 Nanotechnology and biosensors *Biotechnol. Adv.* **22** 505–18
[27] Zhang X, Guo Q and Cui D 2009 Recent advances in nanotechnology applied to biosensors *Sensors* **9** 1033–53
[28] Wang Y and Dmochowski I J 2016 An expanded palette of xenon-129 NMR biosensors *Acc. Chem. Res.* **49** 2179–87
[29] Boutin C *et al* 2011 Cell uptake of a biosensor detected by hyperpolarized 129Xe NMR: the transferrin case *Bioorg. Med. Chem.* **19** 4135–43
[30] Luo Y and Alocilja E C 2017 Portable nuclear magnetic resonance biosensor and assay for a highly sensitive and rapid detection of foodborne bacteria in complex matrices *J. Biol. Eng.* **11** 1–8
[31] Kriz C B, Rådevik K and Kriz D 1996 Magnetic permeability measurements in bioanalysis and biosensors *Anal. Chem.* **68** 1966–70
[32] Hofer C J *et al* 2015 Stable dispersions of azide functionalized ferromagnetic metal nanoparticles *Chem. Commun.* **51** 1826–9
[33] Metaxas P J *et al* 2015 Sensing magnetic nanoparticles using nano-confined ferromagnetic resonances in a magnonic crystal *Appl. Phys. Lett.* **106** 232406
[34] Guo L, Yang Z, Zhi S, Feng Z, Lei C and Zhou Y 2018 A sensitive and innovative detection method for rapid C-reactive proteins analysis based on a micro-fluxgate sensor system *PLoS One* **13** e0194631
[35] Sun X, Feng Z, Zhi S, Lei C, Zhang D and Zhou Y 2017 An integrated microfluidic system using a micro-fluxgate and micro spiral coil for magnetic microbeads trapping and detecting *Sci. Rep.* **7** 1–8
[36] Duyn J H and Schenck J 2017 Contributions to magnetic susceptibility of brain tissue *NMR Biomed.* **30** e3546
[37] Wei H *et al* 2017 Investigating magnetic susceptibility of human knee joint at 7 Tesla *Magn. Reson. Med.* **78** 1933–43
[38] Kim C-B, Lim E-G, Shin S W, Krause H J and Hong H 2016 Magnetic immunoassay platform based on the planar frequency mixing magnetic technique *Biosens. Bioelectron.* **83** 293–9
[39] Achtsnicht S *et al* 2019 Sensitive and rapid detection of cholera toxin subunit B using magnetic frequency mixing detection *PLoS One* **14** e0219356
[40] Lee J, Takemura K and Park E Y 2018 Plasmonic/magnetic graphene-based magnetofluoro-immunosensing platform for virus detection *Sensors Actuators* B **276** 254–61
[41] Singhal C, Dubey A, Mathur A, Pundir C and Narang J 2018 Paper based DNA biosensor for detection of chikungunya virus using gold shells coated magnetic nanocubes *Process Biochem.* **74** 35–42
[42] Drosten C, Doerr H W, Lim W, Stöhr K and Niedrig M 2004 SARS molecular detection external quality assurance *Emerg. Infect. Dis.* **10** 2200–3

[43] Seo G *et al* 2020 Rapid detection of COVID-19 causative virus (SARS-CoV-2) in human nasopharyngeal swab specimens using field-effect transistor-based biosensor *ACS Nano.* **14** 5135–42

[44] Cobo F 2014 Viral hepatitis *Imported Infectious Diseases* ed F Cobo (Cambridge: Woodhead Publishing) pp 213–25

[45] Choi J, Gani A W, Bechstein D J B, Lee J-R, Utz P J and Wang S X 2016 Portable, one-step, and rapid GMR biosensor platform with smartphone interface *Biosens. Bioelectron.* **85** 1–7

[46] Sharma P P *et al* 2017 Integrated platform for detecting pathogenic DNA via magnetic tunneling junction-based biosensors *Sensors Actuators* B **242** 280–7

[47] Altay C, Senay R H, Eksin E, Congur G, Erdem A and Akgol S 2017 Development of amino functionalized carbon coated magnetic nanoparticles and their application to electrochemical detection of hybridization of nucleic acids *Talanta* **164** 175–82

[48] Micheli L *et al* 2017 Towards an immunoanalytical systems for hepatitis a virus determination *Proc. Technol.* **27** 85–6

[49] Alizadeh N, Hallaj R and Salimi A 2017 A highly sensitive electrochemical immunosensor for hepatitis B virus surface antigen detection based on Hemin/G-quadruplex horseradish peroxidase-mimicking DNAzyme-signal amplification *Biosens. Bioelectron.* **94** 184–92

[50] Mashhadizadeh M H and Talemi R P 2016 Synergistic effect of magnetite and gold nanoparticles onto the response of a label-free impedimetric hepatitis B virus DNA biosensor *Mater. Sci. Eng.* C **59** 773–81

[51] Huang R, Xi Z, Deng Y and He N 2016 Fluorescence based aptasensors for the determination of hepatitis B virus e antigen *Sci. Rep.* **6** 31103

[52] Ghafary Z, Hallaj R, Salimi A and Akhtari K 2019 A novel immunosensing method based on the capture and enzymatic release of sandwich-type covalently conjugated thionine–gold nanoparticles as a new fluorescence label used for ultrasensitive detection of hepatitis B virus surface antigen *ACS Omega.* **4** 15323–36

[53] Yang N, Huang Y, Ding G and Fan A 2019 *In situ* generation of prussian blue with potassium ferrocyanide to improve the sensitivity of chemiluminescence immunoassay using magnetic nanoparticles as label *Anal. Chem.* **91** 4906–12

[54] Peters P J, Marston B J, De and Cock K M 2014 HIV: epidemiology in the tropics *Manson's Tropical Infectious Diseases* ed J Farrar, P J Hotez, T Junghanss, G Kang, D Lalloo and N J White 23rd edn (London: W.B. Saunders) pp 68–78

[55] Low H Z *et al* 2016 TLR8 regulation of LILRA3 in monocytes is abrogated in human immunodeficiency virus infection and correlates to CD4 counts and virus loads *Retrovirology.* **13** 15

[56] Phan J C, Nehilla B J, Srinivasan S, Coombs R W, Woodrow K A and Lai J J 2016 Human Immunodeficiency Virus (HIV) separation and enrichment via the combination of antiviral lectin recognition and a thermoresponsive reagent system *Pharm. Res.* **33** 2411–20

[57] Damhorst G L, Kooiman J M and Bashir R 2016 HIV-1 IIIB capture from whole blood on magnetic microparticles *2016 38th Annual Int. Conf. of the IEEE Engineering in Medicine and Biology Society (EMBC)* 16–20 August 2016

[58] Knowlton S, Joshi A, Syrrist P, Coskun A F and Tasoglu S 2017 3D-printed smartphone-based point of care tool for fluorescence- and magnetophoresis-based cytometry *Lab Chip* **17** 2839–51

[59] Bray M and Holbrook M 2019 Hemorrhagic fever viruses ed T M Schmidt *Encyclopedia of Microbiology* 4th edn (Oxford: Academic) pp 525–39

[60] Carinelli S, Kühnemund M, Nilsson M and Pividori M I 2017 Yoctomole electrochemical genosensing of Ebola virus cDNA by rolling circle and circle to circle amplification *Biosens. Bioelectron.* **93** 65–71

[61] Rosenstierne M W *et al* 2016 Rapid bedside inactivation of ebola virus for safe nucleic acid tests *J. Clin. Microbiol.* **54** 2521–9

[62] Zhao J, Fang S, Liu Y, Zeng L and He Z 2020 A lateral flow biosensor based on gold nanoparticles detects four hemorrhagic fever viruses *Anal. Methods* **12** 5613–20

[63] Catenacci L S and De Alcantara B N 2019 Zika virus: a real threat to wildlife? *Fowler's Zoo and Wild Animal Medicine Current Therapy* ed R E Miller, N Lamberski and P P Callevol 9 (London: W.B. Saunders) pp 280–6

[64] Tian B *et al* 2016 Attomolar Zika virus oligonucleotide detection based on loop-mediated isothermal amplification and AC susceptometry *Biosens. Bioelectron.* **86** 420–5

[65] Michelson Y, Lustig Y, Avivi S, Schwartz E and Danielli A 2018 Highly sensitive and specific Zika virus serological assays using a magnetic modulation biosensing system *J. Infect. Dis.* **219** 1035–43

[66] Stern M, Cohen M and Danielli A 2019 Configuration and design of electromagnets for rapid and precise manipulation of magnetic beads in biosensing applications *Micromachines.* **10** 784

[67] Yates M V 2014 Norovirus *Microbiology of Waterborne Diseases* ed S L Percival, M V Yates, D W Williams, R M Chalmers and N F Gray 2nd edn (London: Academic Press) pp 515–22

[68] Lee J, Takemura K, Kato C N, Suzuki T and Park E Y 2017 Binary nanoparticle graphene hybrid structure-based highly sensitive biosensing platform for norovirus-like particle detection *ACS Appl. Mater. Interfaces* **9** 27298–304

[69] Yasuura M, Shirato H, Higo-Moriguchi K and Fujimaki M 2019 Detection of norovirus-like particles with an external force-assisted near-field illumination biosensor *Jpn J. Appl. Phys.* **58** 071005

[70] Davis L E 2014 Influenza virus *Encyclopedia of the Neurological Sciences* ed M J Aminoff and R B Daroff 2nd edn (Oxford: Academic) pp 695–7

[71] Chen H-W *et al* 2017 Targeting and enrichment of viral pathogen by cell membrane cloaked magnetic nanoparticles for enhanced detection *ACS Appl. Mater. Interfaces* **9** 39953–61

[72] Jeun M, Park S, Lee H and Lee K H 2016 Highly sensitive detection of protein biomarkers via nuclear magnetic resonance biosensor with magnetically engineered nanoferrite particles *Int. J. Nanomed.* **11** 5497–503

[73] Haghighi A H, Khorasani M T, Faghih Z and Farjadian F 2020 Effects of different quantities of antibody conjugated with magnetic nanoparticles on cell separation efficiency *Heliyon.* **6** e03677

[74] Mun S and Choi S-J 2015 Detection of *Salmonella typhimurium* by antibody/enzyme-conjugated magnetic nanoparticles *BioChip J.* **9** 10–5

IOP Publishing

Biosensors for Virus Detection

Adil Denizli and Yeşeren Saylan

Chapter 8

Micromechanical biosensors for virus detection

Kutay Sagdic, Özgecan Erdem, Esma Derin, Saeedreza Zeibi Shirejini, Yusuf Aslan, Suleyman Celik, Umit Celik and Fatih Inci

The rapid identification of viruses has vital implications for managing health care, such as controlling disease at early stages and preventing the potential spread of disease. Conventional methods are, however, time-consuming and they often require expensive and sophisticated infrastructure that limits their use in point-of-care settings. Biosensors have attracted notable attention in the last decades, as they are central to applications that monitor many processes, and they offer portability and ease-of-use, enhancing their utility. Among these, micromechanical biosensors rely on small mechanical transducers to sense any vibrations that are precisely proportional to molecular bindings. In this chapter, we comprehensively summarize the field of micromechanical biosensors along with their fundamentals, fabrication, and fundamental characteristics along with pros and cons. Moreover, we mainly focus on how micromechanical biosensors can be tailored to detect viruses such as coronavirus, human immunodeficiency virus (HIV), hepatitis, noroviruses, Ebola, and Zika viruses that pose an overwhelming threat to global health.

8.1 Fundamentals of micromechanical biosensors and actuators

Biosensors reliant on micromechanical resonators are widely utilized to detect specific analytes in liquid, air, or *in vacuo*. These sensors include cantilevers and various other structures such as mechanical components that create acoustic waves and surface stress. In this section, we briefly explain the types and fundamentals of micromechanical biosensors.

8.1.1 Micro-nanomechanical sensors

The development of the atomic force microscope (AFM) began with the use of a cantilevered sharp tip to mechanically probe a surface [1]. In AFM, cantilevers are employed in either static mode or dynamic mode. In the contact mode, the surface is mapped by using the static deviation caused by the contact with the surface as a

doi:10.1088/978-0-7503-3867-7ch8 © IOP Publishing Ltd 2021

feedback source. In the dynamic mode, the cantilever is driven at a frequency close to the resonance frequency, and the surface is therefore mapped using the cantilever vibration amplitude as a feedback source. Later, the flexibility of this technique was demonstrated by the development of several new sensing modes, for example, magnetic properties, capacitance, conductivity, or phase-based measurements.

Recently, micro- and nanometer-scale cantilevers have been studied as sensors using physical principles, which are similar to those found in AFM systems. As basically described above, these cantilevers are usually operated either in the static deflection mode, in which binding on one side of a cantilever leads to unbalanced surface stress, resulting in a measurable up or down deflection, or in the dynamic mode, in which binding on the cantilever increases the mass, and thus decreases the resonant frequency. Cantilever-based sensors offer highly versatile approaches using multiple modes, such as electrostatic, optical, mechanical, and electromagnetic strategies for actuating or sensing cantilever motion to detect chemicals, gases or biological entities.

With the development of microfabrication technologies, significant advances have been made in micro-electromechanical systems/nano-electromechanical systems (MEMS/NEMS)-based biosensor technologies in recent years. MEMS/NEMS cantilever-based sensor applications offer more portable, compact, ultrasensitive options and lower detection limits compared to standard approaches. In addition to their inexpensive features, they provide measurements in real-time, and with appropriate chemical functionalization, they can be employed to monitor minute changes in biological conditions and/or chemical reactions. Moreover, microfabricated devices can be designed with different sizes and shapes and formed into arrays with large numbers of elements. This flexibility of design and size has created the opportunity to combine such devices with microfluidics and miniaturized lab-on-a-chip formats. They are able to monitor small changes in signals caused by molecular/biological vibrations and can be operated with very small sample volumes. Moreover, the relevance of these platforms to the field of virus detection is immediate since they are able to detect biological targets at nanometer-scale [2]. High-frequency MEMS/NEMS platforms are being considered as a new direction in the field of sensing. In particular, the flexibility to adapt their surface properties using specific biomolecules is used to create chemically functionalized MEMS/NEMS cantilevers for biomolecular interaction studies. These devices also exhibit superior potential for multiplexed biosensor arrays due to their small physical dimensions and excellent sensitivity [3, 4]

8.1.2 Cantilever-based micromechanical biosensors

Considering a biosensing approach, the cantilever surface is first functionalized to create a bioreceptor layer, which has a high affinity for the target analytes. When a molecular interaction occurs between the bioreceptor layer and the target analytes, the cantilever surface energy changes in response to the binding of the target to the sensor's surface. As a result, the cantilever deviates statically. In a resonating cantilever biosensor, the resonance frequency alters according to the mass increase or decrease on the cantilever. Moreover, cantilever-based biological or chemical detection methods can be divided into static cantilever bending and resonating cantilever-based detection [5].

8.1.2.1 Static cantilevers

Static cantilevers measure bending or static deflection (static mode: DC mode), which can be induced by external forces, such as gravitational, magnetic, and electrostatic forces exerted on the cantilever or by intrinsic mechanical stresses generated within the cantilever or on the cantilever surface as a result of the analyte interactions [6]. Examining the degree of cantilever bending eliminates the difficulty of viscous damping for experiments performed in a microfluidic chamber that apply a laminar fluid flow around a cantilever, particularly for large analyte sample volumes, reducing the noise generated by fluid turbulence [7]. Isotropic cantilever materials tend to either decrease the surface area (tensile stress) or increase it (compressive stress) as a result of uniform surface stress (σ_s), as we will discuss later. It is worth mentioning that if a cantilever fails to compensate for surface stress equally, the structure can be permanently bent at the opposite sides of a beam or thin plate.

In the static cantilever model, main readout mechanisms include piezoelectric, optical beam deviation or interferometric deviation measurement, and a variety of amplifiers are employed according to the technique. As an interesting example of the static cantilever model, the readout circuitry employs an array of four cantilevers connected to the amplifiers through an analog multiplexer [8]. Moreover, there is a non-aligned cantilever plane that is not deformed between the locations of tensile and compressive stress. Due to the bending deformation, a force F acts at a distance x, resulting in a bending moment $M = F.x$; hence, using the moment of inertia I, and Young's modulus E, the cantilever's radial curvature can be calculated from equation (8.1) [9];

$$\frac{1}{R} = \frac{d^2z}{dx^2} = \frac{M}{EI} \tag{8.1}$$

In the static mode, receptor target binding at the molecular level can be converted into an electrical signal via optical beam deflection, interferometric, piezoresistive, piezo-electric, capacitive, and electron tunneling readout approaches [10–12]. An analysis of the sensitivities reported to date, including the resonance frequencies at which these sensitivities are measured, has been reported in detail by Stemmer *et al* [13]. Any well-designed detection system can achieve a performance such that the shot noise of the electro-optic system is the lowest possible noise background. The detection system noise level determines the measurement sensitivity. Moreover, microcantilever-based sensing platforms can be used in atomic force-based surface imaging to obtain the surface topography of samples from the sensing tip of an AFM [14].

Bicell or quadrant photodiodes are the most used sensors in optical deflection (OBD) sensor systems. In the OBD method, a beam of light (usually laser light) is focused on the cantilever end, and the light is reflected onto the photodiodes. Deviation in the cantilever causes a change in photocurrents in each cell of a quadrant photodiode. Arithmetic differences of photocurrents between quadrants are proportional to cantilever deviation. Nowadays, using this method, cantilever static deviation can be measured with a high precision, and it is successfully used in the resonating mode at bandwidths up to the MHz scale.

A myriad of detection and sensing applications have been developed for the other biological entities, including DNA, RNA, proteins, metals, antibodies, pesticides, antigens, cancer biomarkers, and explosives [15–24]. From the virus sensing and detection perspectives, a triangular microcantilever sensor was developed to detect cantilever-based mass and liquid-borne particle sampling with a resonant frequency of 271 kHz, a sensitivity of 0.13 Hz pg^{-1}, and along with the Q (quality) factor of ~3000. The setup achieved a minimum detectable mass of ~0.05 pg, and can be used to monitor liquid-borne viruses (e.g. coronavirus) (figure 8.1) [25].

8.1.2.2 Resonating cantilevers

Resonating cantilevers are designed to vibrate at a particular resonant frequency. The main materials used in their fabrication are a range of single-crystal materials with micron-scale dimensions. Mainly, the vibration signal of cantilevers is converted to electrical signals by optical beam deflection, interferometric and piezoresistive sensing methods, due to higher achievable bandwidths, greater measurement accuracy, and ease-of-interfacing. Resonating cantilevers may be triggered with mechanical or acoustic waves generated by piezoelectric materials, as well as magnetic, electrostatic, or optical stimuli. Frequency shifts in the cantilever can be detected by phase-locked-loop systems. The frequency of the resonating system can be determined as a function of the measurand, which alters the shape or mass of the resonator by chemically sensitive material coupling or by depositing material onto the resonator. The other components are an amplifier, a detection module, and a vibration drive; vibrations are excited, then sensed in the detection module, and fed back to the drive mechanism via the amplifier. Driving

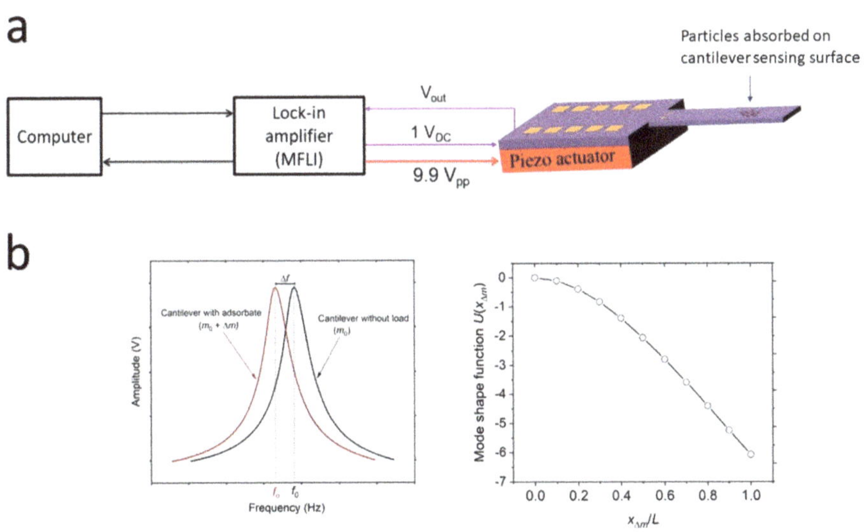

Figure 8.1. (a) A schematic setup for cantilever measurement: a lock-in-amplifier (MFLI) supplies voltage to the in-plane piezo actuator and Wheatstone bridge of a piezoresistive silicon-based cantilever sensor. (b) (i) The frequency shift (Δf) observed in the resonance of a cantilever following exposure to an adsorbate (Δm). (ii) When a uniform beam vibrates under a continuous force at a natural frequency, the deformation data can be shown as the mode shape function, which is obtained at different points along the cantilever. Reproduced from [25], with permission from MDPI. CC BY 4.0.

the efficiency of the vibration excitation and detection mechanisms can be enhanced by modifying the location and the geometry of the electrodes [26, 27].

The resonator design characteristics can be determined using the Q factor (Q). It can be defined using equations (8.2) and (8.3) [27]:

$$Q = 2\pi\left(\frac{E_{\text{stored}}}{E_{\text{lost}}}\right) \tag{8.2}$$

$$Q = \frac{f_0}{\Delta f} \tag{8.3}$$

where E_{stored} is the stored total energy, and E_{lost} is the energy lost per cycle because of damping effects. Moreover, f_0 corresponds to the resonant frequency with the maximum amplitude, and Δf represents the resonance difference of vibrations that are 3 dB lower than the maximum amplitude. The frequency shift measurement accuracy is proportional to the Q factor, and the higher Q factor corresponds to a higher detection limit. On the other hand, the coupling of the resonator results in a reduction in its resonant frequency since exposure to surrounding liquids or gases increases the effective inertia of the resonator [26, 28].

To explain this condition, a myriad of models have been established, including flexural, in-plane, torsional, and extensional oscillations for low-order modes of cantilever beams using two-dimensional analysis [29]. These studies show us that fluid viscosity has a crucial role in the frequency response, especially for two-dimensional flow fields, since the cantilever beams have a high aspect ratio (length/width) [30]. For the 3D forces, fluid compressibility is a crucial parameter affecting the flexural modes for arbitrary mode orders [31].

The sensing abilities of resonating cantilever-based biosensors offer label-free quantitative measurement of biological entities, including viruses, proteins, cells, and nucleic acids using a variety of interfacing electronics and mechanical structures [29, 32]. In particular, silicon-based materials are favored thanks to their low dissipation and high Q factor. Moreover, polymers with high moduli, piezoelectric and magnetoelastic properties have been used as cantilever fabrication materials [33–35]. The fluid cell structure, the inlet and outlet ports are organized in such a way that they are less affected by the noise due to the flow turbulence. Likewise, special immobilization and cleaning processes prevent unwanted behaviors of the system from occurring. In addition, many designs have been studied for single- or multiple layers with either uniform or non-uniform surfaces at appropriate length scales. The surface functionalization on the biosensor surfaces used for biorecognition processes can be accomplished using three steps: (i) surface chemistry preparation, immobilization, and passivation. Subsequently, two biosensing methods can be applied, either continuous measurement of the resonant frequency or measurement of the extremes of the resonant frequencies. For example, an electronic system for a transducer based on a lateral resonating cantilever was developed to predict the dynamic and static behaviors of an electrostatically coupled cantilever-driver structure (CDS) using AC and DC voltages from a driver electrode placed closely parallel to the cantilever. For different voltage polarizations, the frequency response, the static deflection, and the oscillation frequency response were calculated by accounting for the 3D effects, including the fringing field contribution and the real bending of the cantilever, which are relevant in CDSs with low width/gap ratios [36]. In another example, a mass-sensitive cantilever resonating in an air–liquid environment was designed to detect resonance frequency, surface tension changes, and fluorescent intensity changes in 20, 40, and 80 μg mL^{-1} concentrations of rabbit anti-mouse IgG antibodies on the sensor surface. The resonating properties of the cantilever were characterized as follows: a 50% larger Q factor, a 62% larger vibrating amplitude, and a signal-to-noise ratio 5.7 times higher than a completely submerged cantilever; the surface tension was also observed between 57 and 71 mN m^{-1} with a slit width of 6 μm [37].

To benchmark the analytical performance of the device/sensor, biological entities such as bacteria, viruses, nanoparticles, or small molecules, are measured the most. Sensors with very high absolute mass sensitivity can allow analytes to be recognized and counted individually. As shown in some studies (figure 8.2), doubly clamped very-high-frequency nanomechanical cantilever systems can accurately weigh as little as xenon atoms with a noise floor limit of detection (LOD) of ~20 zg and a resolution of 7 zg or 7×10^{-21} g [38]. As label-free biosensors, these systems have been applied to the detection of biological samples, such as viruses, single cells, proteins, and individual strands of DNA [3, 39].

8.1.3 Acoustic wave and piezoelectric micromechanical biosensors

Acoustic waves have been used commercially for over half a century in communications, automotive, and environmental applications. Applying radio frequency or alternating current excitation to electrodes on a piezoelectric material generates acoustic waves that spread across a material surface or perpendicularly form into the bulk medium [40]. The acoustic wave sensor, first reported by Sauerbrey *et al* [41] in 1959, demonstrated the applicability of acoustic waves in biochemical sensing using a bulk thickness shear known as a thickness shear mode (TSM), the mode in a quartz crystal microbalance (QCM), to measure the weight of a thin film by detecting its resonant frequency [42]. Acoustic wave sensors have been adapted for biosensing in

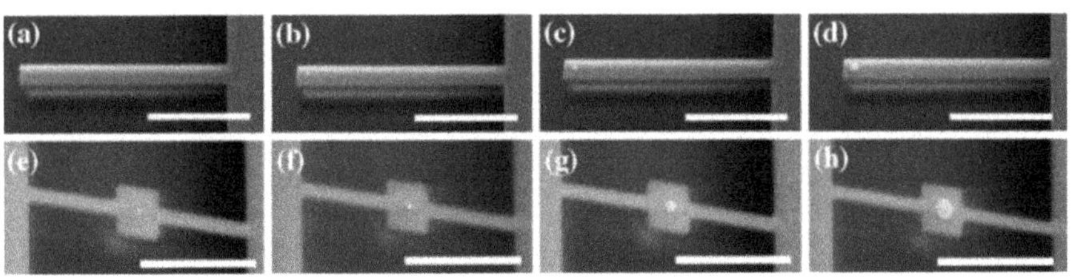

Figure 8.2. A sample of nanomechanical resonators employed in mass sensors with low detection limits. Oblique-angle SEM micrographs of (a)–(d) cantilevers (scale bar = 5 um) and (e)–(h) bridge oscillators (scale bar = 2 um). The diameters of the Au pads from left to right are 50, 100, 200, and 400 nm. Reprinted from [230], with the permission of AIP Publishing.

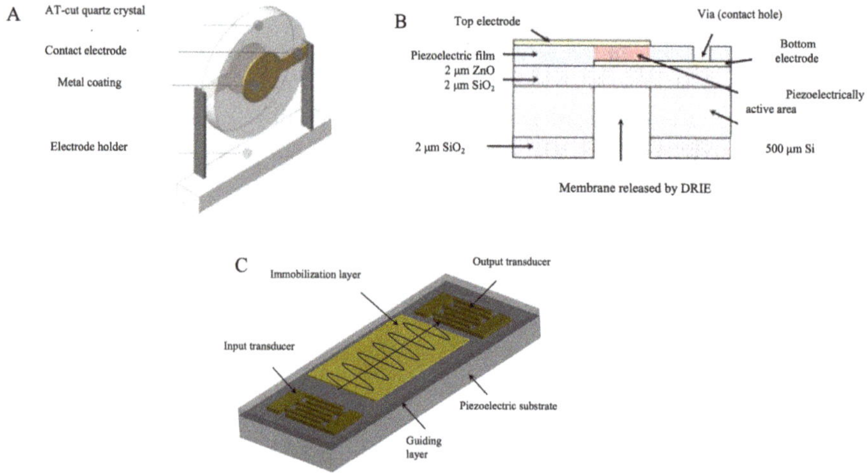

Figure 8.3. A schematic representation of acoustic wave micromechanical sensors. (a) A QCM sensor consisting of an AT-cut quartz crystal coated with two gold electrodes. Reproduced from [231], with permission from MDPI. (b) The schematic cross-sectional view of an FBAR device. Reprinted from [232], Copyright 2012, with permission from Elsevier. (c) A schematic representation of a shear horizontal surface acoustic wave (SH-SAW) device. Reprinted from [233], Copyright 2016, with permission from Elsevier.

recent years due to their high precision, ease of handling, and their ability to process small amounts of biofluids [43]. In addition, acoustic wave-based devices are advantageous because they are less expensive than other sensors without compromising on sensitivity [44, 45]. In acoustic biosensor technologies, acoustic waves pass through a surface formed of a piezoelectric material that has been modified to capture the target biomarker. When binding occurs, the mass of the biomarker changes the properties of the waves, and thus a signal is generated [43]. Using its mass, which is an essential property of any analyte, acoustic/mass-sensitive devices allow analytes to be detected without labels [46].

Acoustic wave sensors are classified according to the waves used: surface or bulk acoustic waves. A bulk wave is described as a wave propagating over a substrate [47, 48]. Two essential devices are reliant on bulk acoustic waves: QCMs and film bulk acoustic resonators (FBARs). The most widely used bulk acoustic wave-based device is the quartz QCM, which has two metallic (usually gold) electrodes separated by a quartz crystal (figure 8.3(a)) [45]. QCM devices use a bulk thickness shear mode that is suitable for liquid and dry environments. These devices commonly have good temperature stability and low bogus bulk wave generation [40]. In QCM biosensors, the crystal surface is frequently covered with a specific receptor that binds to the target analyte. When binding occurs, an increase in the hard crystal mass is observed, which causes a change in the essential resonant frequency of the crystal. This sensor signal is used to accurately determine the mass adsorbed on the crystal surface [46]. The ability of QCM sensors to perform real-time measurements facilitates the possibility of using them in automated continuous monitoring systems. This advantageous feature also makes them applicable for the characterization of affinity-based systems [49]. QCM biosensors are widely used for the detection of bacteria [50, 51], viruses [52, 53], mycotoxins [49], disease markers [54, 55], and many other analytes, even at low concentrations, due to their simple use, low cost, and sensitivity.

The essential operating principle of the FBAR, another bulk acoustic wave-based device, is technically similar to those of QCM devices. FBAR devices have a smaller base mass than those of QCMs, enabling a much higher resonance frequency and eventually providing more sensitivity. Moreover, the operating resonance frequency of FBAR devices is generally significantly higher than those of QCMs (sub-GHz to 10 GHz) [56]. The working principle of FBAR-based mass sensors is reliant on the mass loading effect. Consequently, as the mass loading on the resonator increases, downward shifts occur in the resonant frequency [57]. Unlike QCMs, FBARs employ piezoelectric film instead of quartz crystal as the resonance unit. Hence, they can be developed at small sizes, allowing higher sensitivity detection [58]. FBAR devices typically have a three-layer structure, as shown in figure 8.3(b), consisting of a lower electrode, a piezoelectric layer, and an upper electrode sputtered onto a silicon substrate. When an alternating current signal is applied to the piezoelectric material, the electrical signal is converted into an acoustic/mechanical wave that spreads lengthwise in FBAR devices [57].

Devices based on surface acoustic waves (SAWs) were first designed in 1965 [59], and these devices were utilizing interdigital transducers to generate SAWs. Subsequently, extensive research into SAW sensors for a wide variety of applications began to increase rapidly [60]. Among acoustic devices, SAW devices stand out due to their small size, extremely high frequency, rapid response, high durability, and integration with various receiver materials. These properties make them highly suitable for the detection of chemical warfare agents [61], volatile organic compounds [62], toxic gases and explosives [63, 64], and biological analytes [65, 66]. These devices are made of piezoelectric materials, on which a periodic comb-shaped interdigital transducer (IDT) pattern is created using photolithography or another process. After an AC voltage is applied to an IDT, an acoustic wave propagates that spreads across the crystal surface [67]. SAW devices are grouped into three categories according to the type of surface wave: horizontal shift/Love wave sensors, Lamb wave (L-SAW) sensors, and Rayleigh wave (R-SAW) sensors [68].

Shear horizontal surface acoustic wave biosensors provide precise and cost-effective real-time detection of many analytes and biomolecules without labels, using horizontally polarized surface shear waves. These biosensors typically comprise a planar electrode formed from a piezoelectric substrate that contains IDTs (figure 8.3(c)). Lithium tantalate, meeting most of the requirements for efficacious horizontally polarized surface shear wave production, is employed as a substrate. These waves are produced by IDTs using an AC voltage at a high frequency from 80 to 400 MHz [69]. The shear horizontal wave generated in these devices has a displacement parallel to the substrate. In-plane propagation provides the minimum damping of waves in environments in which the detection area is liquid. Therefore, shear horizontal surface acoustic wave biosensor devices are more suitable for research aimed at detection in liquid environments [70]. Another type of SH-SAW device that is suitable for SAW biosensors is the Love wave sensor, which is defined as the acoustic resonance of an accumulated layer in which horizontally polarized surface waves move across a substrate [71]. Love wave-type surface acoustic wave sensors are one of the latest designed devices for the incorporation into lab-on-a-chip systems, and they are especially appropriate for DNA, antibody, and protein analyses. Love waves spread to a close vicinity of a piezoelectric material's surface, which supplies horizontal shear waves if placed above a piezoelectric substrate with a lower shear rate. The amplitude and the speed of the wave mainly depend on changes in the near-surface media [72].

Rayleigh waves are the other type of surface elastic wave of the broad acoustic wave family, and they are similar to mechanical waves on the Earth's surface [73]. Rayleigh waves can be defined as guided waves capable of propagating through a semi-infinite solid medium surface. This wave preserves the majority of its energy, compared to other kinds of body waves such as S-waves or P-waves, and it also decays very slowly at the surface when the half-space is subjected to a harmonic load or displacement [74]. R-SAW sensors are commonly used as gas sensors, in which role they are responsive, yet they are not suitable for real-time detection in liquids due to SAW damping. The high sensitivity of R-SAW sensors occurs because their wave energy is naturally limited to the substrate without the need for a guiding layer [75]. On the other hand, Lamb wave devices support two different spreading approaches: symmetrical and anti-symmetrical, which reflect the symmetry of particle motion around the plate's median [76]. Recently, as a good alternative to the operational limitations of SAWs, Lamb wave-based sensors have been designed to work with acoustic flow, sorting, mixing and particle patterning in a sessile droplet [77]. Generally, a Lamb wave is elliptically polarized with two-particle displacement components, namely shear vertical and longitudinal. According to the wavelength ratio of the specific plate thickness, a Lamb wave has a predominantly in-plane longitudinal component with a near-zero shear vertical component [78].

The piezoelectric effect is the phenomenon whereby electrical charge forms on certain materials due to the application of mechanical stress. This effect is a reversible process, meaning that a volume change can be induced by an electric field [79]. Therefore, piezoelectric materials are perfect for the transducer applications. Piezoelectric materials have recently garnered significant attention for sensor applications. They are composed of various substances, such as organic materials (e.g. polylactic acid (PLLA) [80] and polyvinylidene fluoride (PVDF) [81]), inorganic materials (e.g. lead zirconate titanate (PZT) [82], and aluminium nitride (AlN) [83]), composites, and biomaterials (e.g. cellulose [84], collagen [85], and chitin [86]). Additionally, silicon dioxide (SiO_2) [87] are also preferred for their high mechanical durability. Although most organic polymers are already biocompatible, inorganic materials may require encapsulation to achieve biocompatibility [88].

In addition, the use of piezoelectric material as a mass sensor causes the frequency of the system to change; such a change can be calculated from Sauerbrey's equation (equation (8.4)) [40, 89]:

$$\Delta f = \frac{2\Delta m f_0^2}{A\sqrt{\rho\mu}} \tag{8.4}$$

where Δm is the mass loading, f_0 is oscillation frequency, p is the density of the piezoelectric material, A is the surface area and μ represents the ratio of the mass to the area. Additionally, a coupling factor, k^2, is utilized to perform the conversion between electrical energy and mechanical energy (equation (8.5)) [40, 90]:

$$k^2 = \frac{e_{31}^2}{c_{11}\varepsilon_{33}} \tag{8.5}$$

where c_{11}, e_{31}, and ε_{33} represent the electrical field, the elasticity of the material, and absolute permittivity, respectively. The coupling factor value depends on the type of piezoelectric material used, the composition of its guiding layers, and the thicknesses of the layers.

The sensing of biological and chemical markers on a deposited piezoelectric surface is achieved by analyzing changes in the mechanical stress, electrical charge, and applied mass. For instance, piezoresistors are utilized with MEMS to sense low concentrations of markers by measuring the surface stress of the material. Temperature is the main drawback of this method. However, utilizing four piezoresistors as a Wheatstone bridge may overcome this issue [91].

Strain caused by periodic external stress leads to wave propagation inside piezoelectric materials. This longitudinal or transversal wave defines the frequency range of the device [89]. The wave spectrum utilized by piezoelectric biosensors consists of many modes, such as the acoustic plate mode (APM), thickness shear mode (TSM), SAWs, and the flexural plate wave (FPW) mode [40]. For instance, the FPW mode is a vibrating mode that is induced by a substance composed of piezoelectric material and structures resembling cantilevers or membranes. The stiffness coefficient of the cantilever is changed by the binding of molecules to the functionalized surface. This affects the resonance of the sensor, since the bending moment on the cantilever is altered, and markers are detected on the surface [92]. In a study, FPW devices were integrated with self-assembled monolayers (SAMs) followed by the immobilization of immunoglobulin-E (IgE) antibodies for allergen detection [93]. Additionally, piezoelectric materials are new alternatives to conventional semiconductor materials in micromachined ultrasonic transducers (MUTs). These devices are especially promising for medical diagnosis, fingerprint recognition, and implant power supplies [94].

Immunosensors composed of piezoelectric materials detect specific antigen groups and measure their concentrations using immobilized antibodies on their surface along with favorable surface functionalization. As an antigen attaches to the immobilized antibody, the frequency of the system oscillations alters, and biorecognition occurs. Antigens with high molecular weights are preferred since their increased stimulation causes a more significant decrease in the oscillation frequency. However, this effect can also be obtained by introducing nanoparticles covered with antibodies [95].

8.1.4 Surface stress-based biosensors

The surface stress, σ_s, is the magnitude of work per unit area required to stretch a surface (figure 8.4) [96]. Intermolecular forces arising from adsorbate–surface and

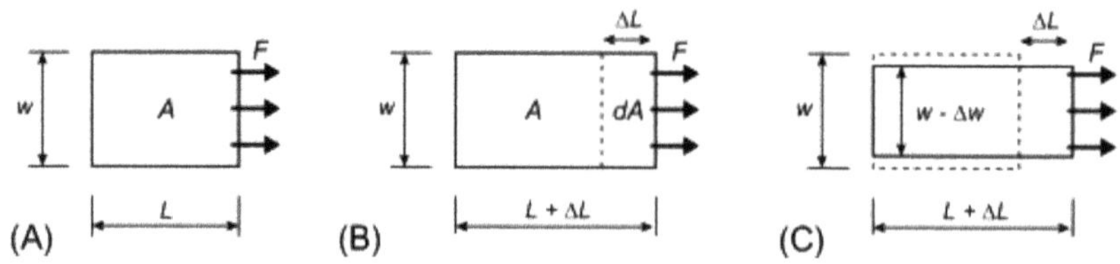

Figure 8.4. Surface stress on a cantilever. (a) A stretching force acting on a surface. (b) Graph of one-dimensional deformation. (c) A design that considers lateral deformation. Reprinted from [96], Copyright 2019, with permission from Elsevier.

adsorbate–adsorbate interactions of small molecules are known to be factors that induce surface stress, which results from the mechanical bending of a solid substrate [97]. The surface stress σ_s and F are given by equations (8.6) and (8.7):

$$\sigma_s = \frac{dW}{dA} = \frac{F}{w} \tag{8.6}$$

$$F = \sigma_s . w \tag{8.7}$$

where w, L, A, and F are the width, length, area, and force, respectively. In this scenario of one-dimensional stretching (see figures 8.4(a) and 8.4(b)), the new area $dA = w.\Delta L$, and the work $dW = F.\Delta L$.

The concept of surface stress has been known since the 1900s, and it was first applied at the end of the 20th century for sensing studies [98–100]. The hypothesis of surface stress is derived from the minuscule repulsive or attractive forces that can be generated on a cantilever surface because of the alterations in its electronic energy states or charge distribution when target–receptor interactions occur [101]. Surface stress applied to a cantilever can be described by Stoney's formula using different dimensions (force per length) when the film thickness is very small or it obeys an unknown function (8.8) [102]:

$$\sigma_s = \frac{Et^2}{6R} \tag{8.8}$$

where E denotes Young's modulus, t represents thickness, and R is the radius of curvature resulting from the interactions of molecules with the thin-film cantilever. Furthermore, the deflection, δ of a cantilever system can be explained by the bending obtained from the radius R and the length, L, of the cantilever curvature equation (8.9) [96, 102]:

$$\delta = \frac{3L^2}{Et^2}\sigma_s \tag{8.9}$$

From a chemistry perspective, it can be explained by an atomic bonding at the surface. If bond strengths between surface atoms are more powerful than those between subsurface atoms, tensile surface stress is generated by attractive forces, resulting in a concave surface curvature. On the other hand, if the surface atoms tend to repel each other, compressive surface stress is induced, resulting in a convex surface curvature [103, 104].

In the early developmental stage of surface stress-based biosensors, scientists made a good analogy between biology, in which the interactions of membrane molecules result in the lateral tension of a lipid bilayer and a cantilever system [105, 106]. This technology includes a relatively new class of biosensor, which encompasses three main features: (i) a transporter of analytes, (ii) a sensitive element that has a selective layer and a cantilever or membrane as a transducer, and (iii) a signal detection system [104]. Such biosensors can be described as microscale and label-free systems that work in the reported range of 1 mJ m^{-2} to 900 mJ m^{-2} [97, 107–110].

The transporter of analytes is geometrically restricted to a small size to meet the requirements for microdomain forces and low energy consumption. Microfluidics is generally employed to mimic conventional analyte handling techniques, which enables the use of facile fabricated, highly integrated systems, allowing the rapid analysis of biological materials [111]. The sensitive element of such sensors highlights the ability of surface stress sensors with selective layers to detect a myriad of reactions, e.g., antigen–antibody binding [112], DNA hybridization [113], SAM formation [114], and phase-transitive reactions [115]. The transducer layer can either be a membrane or a cantilever. This element basically has two main layers: a submaterial layer (such as silicon, aluminum nitride, silicon nitride, polymethylmethacrylate (PMMA) or polydimethylsiloxane (PDMS)) and a contact layer, i.e. a gold layer. For instance, a microcantilever-based aptasensor was developed with thrombin DNA aptamer (TBA15–1) as a biorecognition element using a plasma treatment from the top, and a layer of gold was coated from the bottom side and developed using an SU-8 polymer and MEMS lithography [116]. Compared to its silicon counterparts, the deflection value of the SU-8 polymer-based microcantilever was found to be an order of magnitude greater, due to its small Young's modulus. However, cantilevers are mostly developed stiff substrates with low selectivity and anisotropy, resulting in very limited sensitivity, yet good biocompatibility. Moreover, cantilever geometry is not the best option for sensing in liquid media, and thus, it can severely reduce the signal-to-noise ratio (SNR) because of non-specific adsorption on the backside of the cantilever [117]. Hence, membranes are more convenient than cantilevers for sample isolation and detection. For example, a surface stress-based capacitive biosensor was developed via the structural optimization of a micromembrane consisting of a PDMS submembrane coated with gold that served as a sensitivity layer. The membrane shape, the coverage of gold, and the thermal strain mismatch were optimized using deflection and capacitance finite element analysis (FEA, ANSYS®). The sensor responses to reactions causing deflection for a surface stress change σ_s and the equivalent temperature load ΔT were calculated using equations (8.10) and (8.11), respectively [118]:

$$\frac{1}{R} = 6\left(\frac{1 - v}{Et^2}\right)\sigma_s \tag{8.10}$$

$$\Delta T = \frac{\sigma_s}{Eg.\ \partial g.\ Tg} \tag{8.11}$$

where v is Poisson's ratio, E_g is Young's modulus, ∂_g is the coefficient of thermal expansion, and T_g is the thickness of gold used. The results demonstrated that the geometry and gold coverage of the membrane affects the deflection rate and the sensitivity of the sensor system.

Signal detection setups can generally be classified into those based on electrical detection and optical detection. The optical systems can be categorized into those based on radioactive [119], fluorescent [120], and fiber interferometry detection [121]. The electrical systems can be further classified into those based on piezoresistive [122] and capacitive detection [123]. In principle, microcantilever approaches depend on easy and rapid optical methods to detect deflections without any internal sensing material that can disturb cell environment. On the other hand, the thermal effect of heat drift constitute a disadvantage that causes an excessive temperature increase in the cantilevers [124]. From an application perspective, a surface stress biosensor utilizing a MEMS-based Fabry–Perot interferometer for label-free sensing of an antigen–antibody interaction was designed on a movable parylene C membrane with 300 nm of Au film. By editing the spring constant and coverage ratio of the Au half-mirrors, the sensitivity and selectivity were modified to detect the amount of deflection. The wavelength selectivity was improved by 6.6 times through integrating these mirrors further to that a reflection peak shift since bovine serum albumin (BSA) antigen-antibody interaction was observed spectroscopically using the fabricated optical interferometer [125]. Compared to optical detection, electrical detection is a relatively low-cost and facile system that can be used with opaque solutions, yet its fabrication process is complex, and the stiffness of the materials used is larger than those of the usual sensing elements, thereby reducing the sensitivity. From an application perspective, a piezoresistive membrane-type nanomechanical cantilever sensor created by boron diffusion onto single-crystal silicon has emerged for the label-free and real-time detection of analyte-induced surface stress. The adsorbate membrane consists of four piezoresistive sensing beams, consisting of a full Wheatstone bridge, which is ~43 times more efficient than a standard cantilever (figure 8.5). The relative resistance change ($\Delta R/R$) was evaluated by FEA (COMSOL Multiphysics 3.5a) by applying compressive surface stress ($-3.0\ \mathrm{N\ m^{-1}}$), resulting in downward deflection and tensile stresses in the x (σ_x) and y (σ_y) directions, where π_{44} ($\sim 138.1 \times 10^{-11}\ \mathrm{Pa^{-1}}$) is a typical piezoresistance coefficient in the equation as follows (8.12) [126]:

$$\frac{\Delta R}{R} = \frac{1}{2}\pi_{44}(\sigma_x - \sigma_y). \tag{8.12}$$

8.2 Virus detection using micromechanical biosensors

In history, we have faced many clinical conditions and diseases; however, infectious diseases have gained the attention of researchers because they can spread to other individuals in various ways, such as through contact, contaminated air, water or

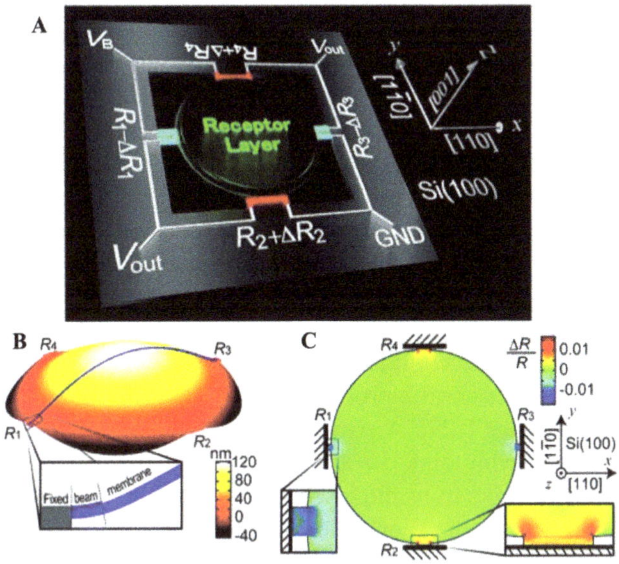

Figure 8.5. (a) A schematic illustration of the membrane-based sensor with p-type piezoresistors on a single crystal of n-type silicon. (b) The deformed shape of the membrane. (c) The distribution of $\Delta R/R$ on the surface of the sensor with current flowing in the x direction. Reprinted with permission from [126]. Copyright 2011 American Chemical Society.

food [127, 128]. Therefore, controlling infectious diseases is crucial to maintaining global health. In addition, socio-economic, environmental, and ecological factors are also strongly influenced by infectious diseases [127]. In the last decade, infection-related mortality and morbidity have increased at an alarming rate. More specifically, the various types of viruses (e.g. HIV, dengue, ebola, Zika, influenza, hepatitis, norovirus, severe acute respiratory syndrome (SARS), and SARS-CoV-2) have a significant impact on multiple aspects in society, science, and technology [129]. Traditional diagnostic procedures are mainly based on enzyme-linked immunosorbent assay (ELISA), nucleic acid detection (e.g. a polymerase chain reaction (PCR)), culturing, or microscopy, which are complex, time-consuming, and expensive, and also they require sophisticated instruments and highly skilled personnel for their proper operations [130]. As a consequence of these facts, simple, rapid, accurate, portable, multiplex, and inexpensive diagnostic platforms are among the primary concerns of research into early detection, which is a crucial clinical step required to initiate effective treatment [128]. Considering the drawbacks of conventional detection techniques, the necessity for a novel detection platform cannot be ignored. Therefore, recent advances in biosensor development offer crucial solutions and provide rapid, cost-effective, and facile platforms. They present notable improvements in sensitivity and multiplexed sensing modalities for point-of-care (POC) settings [130, 131].

8.2.1 Coronavirus

Over the last few decades, three major types of pathogenic coronavirus, i.e. SARS-CoV, Middle East respiratory syndrome coronavirus (MERS-CoV), and SARS-CoV-2, have spread to large territories, resulting in a high rate of morbidity and mortality [132, 133]. SARS-CoV was the first coronavirus that emerged in China in 2002, spreading rapidly to more than 30 countries and leading to about 8000 SARS-CoV cases [134]. Later, in 2012, a new type of coronavirus known as MERS-CoV was discovered in Saudi Arabia. It was found that camels transmitted it to humans, and it spread to over 27 countries, with a total of 858 cases recorded worldwide [135]. The latter, SARS-CoV-2, was detected in China in December 2019. The impact of the COVID-19 pandemic continues all over the world [136]. This virus spreads rapidly, and it causes serious health problems, such as difficulty in breathing, limb pain, fever, and severe coughing [129]. A patient infected with SARS-CoV-2 takes two to seven days to display symptoms, so about half of all patients are asymptomatic. Therefore, it can easily be transmitted from a patient to a healthy person [137]. According to a survey from the World Health Organization (WHO), over 264 million COVID-19 patients have been identified and over 5.2 million people have died up to December 2021.

In order to suppress the spread of the virus, it is essential to monitor and track the COVID-19 status of persons. A wide variety of imaging techniques, including chest computed tomography (CT) scans [138], single-photon emission computed tomography (SPECT), and positron emission tomography (PET) have been utilized for COVID-19 detection. These methods, however, have low sensitivity, and they are bulky and expensive. Moreover, they can only be performed by trained professionals in centralized hospitals [139]. Furthermore, these methods cannot be easily integrated into the POC settings for broad applicability in screening studies [140]. Briefly, current COVID-19 detection methods include ELISA [141], loop-mediated isothermal amplification (LAMP) [142], chemiluminescence [143], CRISPR (clustered regularly interspaced short palindromic repeats) [144], and reverse transcription-polymerase chain reaction (RT-PCR) assay [145]. Antibody-based detection methods, such as ELISA, are not suitable for early diagnosis. Moreover, CRISPR and immunoglobulin G (IgG) can only be detected in the patient's blood after five to ten days. Another common issue in this process is false-positive results caused by the interference of other proteins in human blood [146]. Although RT-PCR is the gold-standard method for SARS-CoV-2 detection, it is laborious, time-consuming, and it requires a trained person to carry out the process, including purification, nucleic acid amplification, and fluorescence detection. Moreover, this method has a high false-negative ratio (FNR), and people who have false-negative results can spread viruses and infect other individuals [147].

On the other hand, biosensors are powerful tools for COVID-19 detection, offering high sensitivity, high accuracy, and fast turnaround. Majorly, biosensors are classified as optical, electrochemical, electrical (FET), and mass-based biosensors [148, 149]. Due to their advantages, piezoelectric biosensors, also known as mass-based mechanical biosensors, have piqued the interest of researchers seeking to

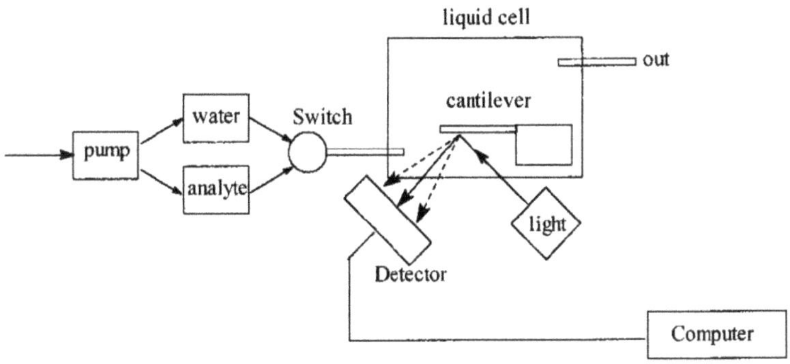

Figure 8.6. The schematic diagram of the instrument used to measure cantilever deflection and demonstrate the glass flow cell. Reprinted with permission from [234]. Copyright 2001 American Chemical Society.

detect COVID-19. In 2004, a QCM piezoelectric immunosensor was developed for SARS-CoV detection from the sputum gas phase. Briefly, horse polyclonal antibodies were immobilized on a crystal surface through protein A in order to achieve an oriented antibody structure. The viruses were specifically adsorbed by the antibodies after an incubation. The mass load on the sensor was changed due to virus adsorption on the crystal surface, leading to frequency variations. The frequency was linearly correlated to the SARS-CoV concentrations, with a dynamic measurement range of 0.4–6 μg mL^{-1}, and an analysis time of less than 2 min. The biosensor exhibited good reproducibility, and the device could be reused 100 times without the loss of activity. In addition, the sensor remained stable for two months while kept over silica gel blue at 4°C – 6°C [150].

In another study, a microcantilever sensor was utilized for feline coronavirus (FIP) detection. A polyclonal type I FIP anti-viral antibody was immobilized on the cantilever surface. The cantilever bending and its deflection were measured using the optical beam deflection method as the sample approached the unit cell of the sensor (figure 8.6). This research achieved a limit of detection of 0.1 μg mL^{-1} and an assay running time of less than an hour. This work provides an opportunity for the further utilization of microcantilever sensors for the detection of human-associated SARS-CoV [151].

Recently, a MEMS biosensor conjugated with PZT piezoelectric discs was developed for COVID-19 detection. The piezo discs were coated with a sialyl glycopolymer cross-linker polymer receptor to increase the induced surface stress and SL-polyacrylamide (SL-PAA) in order to target the particular protein found on the virus for enhancing device specificity and signal amplification. As the sample was injected, the radial mode resonance frequency shifted due to the surface stress caused by the addition of the sample and the viruses. It was demonstrated that the resonance frequency shift and the virus concentration had a linear correlation. It was also shown that as the piezoelectric disc's thickness decreased, the sensitivity increased. In this study, the SARS-COV-2 concentration was measured in the range of 10^4–10^8 virions mL^{-1} [152].

8.2.2 Human immunodeficiency virus

HIV is a virus that attacks the host's immune system and causes acquired immunodeficiency syndrome (AIDS). According to the WHO, 38 million people were living with HIV, 1.7 million new infections occurred and ~1 million people died because of HIV-related disease up to the end of 2019 [153, 154]. These statistics emphasize that HIV infection is a primary public health concern, as it causes high mortality and morbidity. HIV first came to light in the 1980s, when homosexual men were found to have advanced and unexplained immunodeficiency. The origin of the virus is thought to be non-human primates (chimpanzees), and it has spread all over the world over the years [153]. The prevalence of HIV infection in low-income countries (especially in sub-Saharan Africa) is much higher than developed countries. The main reason for this situation is low levels of access to high-throughput systems for diagnosis and treatment. The conventional diagnostic procedures consist of nucleic acid tests (NAT), antigen/antibody tests, and antibody tests using blood, serum or oral fluid samples [153].

In one study, for instance, a microcantilever-based principle was employed to detect HIV envelope glycoprotein (Env) gp120 [155]. Accordingly, the cantilever was functionalized using monoclonal antibodies (mAbs) A32 or T8 for specific binding to gp120, and after HIV was captured, the deflection change was compared to the baseline (figure 8.7(a)–(c)). The developed platform proved that cantilever-based biosensor systems are promising methods for HIV-1 infection detection, offering easy-to-use and portable systems [155]. Another study introduced a platform to detect HIV-2 [156]. This platform utilized a label-free immunosensor approach combined

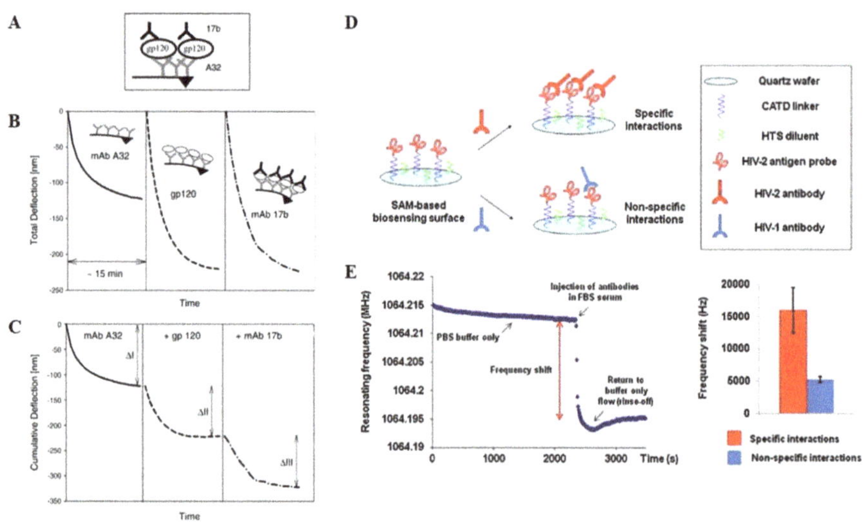

Figure 8.7. (a) A cantilever functionalized with SAMs and antibodies. (b) Total deflection and (c) cumulative deflection changes with respect to time. Reprinted from [155], Copyright 2006, with permission from Elsevier. (d) The illustration of QCM surface functionalization. (e) The frequency change was observed when antibodies were injected into the QCM sensor. Reprinted from [156], Copyright 2011, with permission from Elsevier.

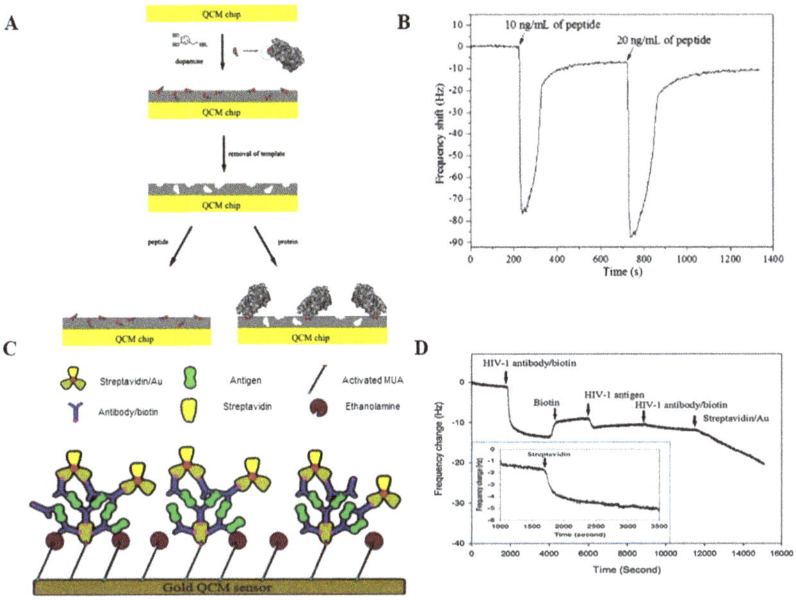

Figure 8.8. (a) The schematic representation of epitope imprinting. (b) The frequency change was obtained when the template was applied in two different concentrations (10 ng mL^{-1} and 20 ng mL^{-1}) and returned to the baseline. Reprinted from [157], Copyright 2012, with permission from Elsevier (c) A schematic representation of surface functionalization with MUA, streptavidin, biotin–streptavidin conjugation, antibody–antigen interactions, and streptavidin–Au binding. (d) The frequency change was obtained due to the applied ligands. Reprinted from [158], Copyright 2016, with permission from Elsevier.

with an electromagnetic piezoelectric acoustic sensor (EMPAS). Herein, a quartz wafer surface was functionalized with a self-assembled monolayer and HIV-2 immunodominant epitopes (see figure 8.7(d)). Specificity and selectivity analyses were performed using HIV-2 or HIV-1 monoclonal antibodies [156]. On the other hand, QCMs are among the most popular micromechanical biosensor designs, since a variety of approaches can be employed. In contrast to the classical approach (e.g. antibody or aptamer functionalization), one of the studies employed molecularly imprinted polymer (MIP)-coated surfaces to create specific grooves for accommodating target molecules [157]. In this study, MIPs were fabricated via epitope imprinting techniques using the HIV-1 gp41 protein template (figure 8.8(a)). Epitope-imprinting polydopamine film not only offers template recognition, but also provides the recognition of the corresponding protein. The use of polydopamine was another novelty of this study, since its use in epitope imprinting has not been common compared to that of dopamine, which is often employed in MIPs due to its high stability and biocompatibility. In addition, facile protocols, high stability, and sensitivity are benefits of these platforms [157]. Another QCM technique was adapted to develop a diagnostic tool for HIV-1 detection [158]. This technique was combined with gold nanoparticles (AuNPs) to enhance the signals that detect the window of the HIV infection period (figure 8.8(c)). Briefly, AuNPs were synthesized by reduction, and streptavidin was then decorated on their surfaces. At the surface chemistry steps, the sensor surface was first treated with 11-mercaptoundecanoicacid (MUA) and streptavidin. Next, biotinylated polyclonal antibodies, which are capable of capturing HIV-1 p24 antigen, were immobilized. This platform is a promising tool for HIV-1 detection from serum samples, and it provides the detection of antigens, as low as 1 ng mL^{-1} [158]. In addition, a piezoelectric biosensor was developed to detect HIV-1 and HIV-2 from human serum. The surface was functionalized with antibodies against HIV-1 and HIV-2. The LOD was found to be 12 tissue culture infective doses (TCID50s) for HIV-1 (approximately 3,000 viral particles) and 87 TCID50s for HIV-2 in buffers. Considering the nucleic acid techniques, this platform offers some advantages in terms of accuracy, the precision of measurement, and operation [159]. In another example, a microcantilever sensor was designed to capture p24 antigen in order to determine HIV infection. The dynamics of the beam were simulated by the COMSOL Multiphysics software to model the stress and displacement. Elastic silicon-based (SiO_2, Si, and Si_3N_4) materials were used as wafers, since Young's modulus and Poisson's ratio play critical roles in the deflection rates of different geometries. The results showed that SiO_2 and a V-shaped design were the optimum parameters, due to larger amounts of deflection [160].

8.2.3 Hepatitis

Hepatitis, i.e. the inflammation of liver, holds serious health issues, leading to cirrhosis and liver cancer. This inflammation is majorly caused by various hepatitis viruses, including hepatitis A (HAV), hepatitis B (HBV), hepatitis C (HCV), hepatitis D (HDV), hepatitis E (HEV), and hepatitis G (HGV). Briefly, HAV infects an estimated 1.4 million people worldwide per year [161]. It has been estimated that two billion people worldwide have been infected with HBV. Moreover, 360 million people are chronically infected by HBV, and 600 000 die each year from HBV-related liver disease [162]. Currently, approximately 170 million people worldwide are affected by HCV, and about 3–4 million people are infected globally by the HCV virus each year [163–165]. Hepatitis viruses can be classified according to their severity, chronicity, genomic types, antigenic patterns, and transmission mode [166]. Considering the vast number of people worldwide who have hepatitis and the scarcity of cures for all forms of hepatitis, early detection of hepatitis viruses in various bodily fluids is crucial for controlling and limiting the spread of these viruses, as well as reducing liver-related disease, cancer, and mortality rates.

Several attempts have been made over the last two decades to detect specific biomarkers located on the surface of the hepatitis viruses or to capture antibodies produced against hepatitis viruses. Immunoassay is one of the most frequently employed serological methods for detecting hepatitis viruses. It focuses on the interactions between antibodies and antigens (biomarkers) [167]. This method can be expanded to ELISA [168], RIA (radio-immunoassay) [169], immuno-chromatographic assay (ICA) [170], immuno-chemiluminescence [171], and lateral flow immunoassay (LFA) [172]. Generally, in the immunoassay method, an antigen is captured by an appropriate antibody and the response can be qualitatively or quantitatively measured. This method's key advantages are low reagent consumption, high sensitivity and stability, low LOD, and ease-of-use. However, the immunoassay approach is expensive and time-consuming, rendering it is unsuitable for clinical applications because the approach needs to be faster and cheaper considering its expansion to the POC and resource-scarce settings [173]. Basically, LFA uses a sample pad, a conjugate pad, an analytical membrane, and a wicking pad. Once the sample is introduced, it passes through the LFA device by capillary force, and later on, reaches the detection site where antigens or antibodies in the sample interact with immobilized antibody antigens on the LFA platform, resulting in a colorful signal that can be distinguishable by a naked-eye [174]. While LFA has higher specificity, sensitivity, and a lower processing time than other immunoassay methods, making it more suitable for clinical applications.

The molecular detection technique is another impactful method used for various types of hepatitis virus detection; in particular, HBV and HCV have been detected with high accuracy as shown in the literature [175]. Molecular approaches include real-time PCR [176], NAT [177], and RT-PCR. In summary, the PCR method relies on the amplification of nucleic acids, mainly RNA and DNA. Accordingly, several DNA target fragments are amplified in a single tube after a large amount of the desired nucleic acid is synthesized in a short period of time. The primary application of NAT methods is to amplify and detect RNA. Molecular detection-based methods have higher sensitivity, specificity, and a wider dynamic range than serological methods. They can also be beneficial when serological techniques fail to respond appropriately [178]. In particular, HBV DNA can be identified using the PCR method even if all serological biomarkers are negative [179]. However, molecular detection methods necessitate complex instruments and post-processing, and they are incapable of demonstrating virus viability [175].

A wide variety of biosensors including electrochemical [180], optical [181], and mass-based (mechanical) sensors [129] have been utilized in the literature for virus detection. Piezoelectric biosensors are prevalent among researchers because of their advantages, which include high sensitivity and specificity, low cost, real-time analysis, and label-free detection [182–184]. A QCM biosensor has been developed for HBV DNA detection [185]. In this research, an HBV nucleic acid probe was immobilized on a gold electrode surface using polyethyleneimine adhesion, the polyethyleneimine glutaraldehyde (PEI-Glu) crosslinking method, and physical adsorption. The results showed that as compared to adsorption immobilization, the use of polyethyleneimine adhesion and glutaraldehyde crosslinking leads to the improvements in reproducibility and stability. Furthermore, HBV DNA concentration was able to be measured in the range of 0.02–0.14 $\mu g\ mL^{-1}$, and the biosensors provided reusability, so that the dynamic range was not affected after five time-uses of the regeneration procedure. In another study, peptide nucleic acid (PNA) rather than DNA was employed for real-time monitoring of HBV DNA-PNA hybridization [186]. The PNA probe was immobilized on a gold surface through biotin-avidin binding, which replaced the traditional DNA probe for the direct detection of HBV genomic DNA without prior amplification by PCR. The

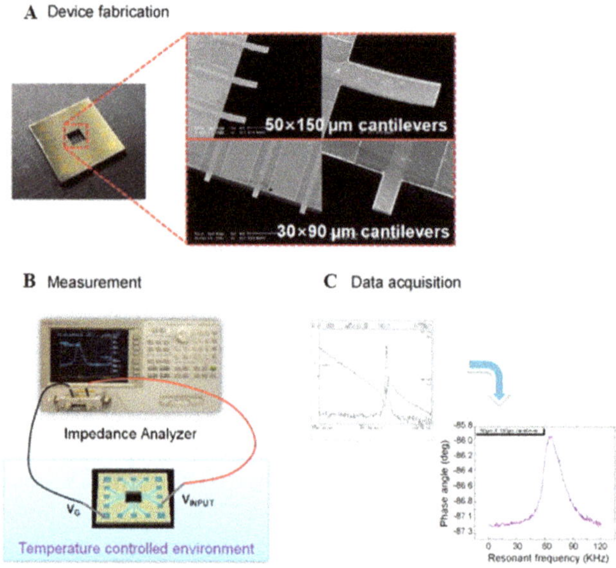

Figure 8.9. Dynamic microcantilever and resonant frequency apparatus measurements. (a) A device composed of twelve microcantilevers and SEM images of two different types of the microcantilever. (b) A schematic representation of an impedance analyzer to determine the resonant frequency. (c) The resonant frequency data was extracted from an impedance signal. Reprinted from [191], Copyright 2009, with permission from Elsevier.

device was relatively fast, and the hybridization assay was completed in 50 min. In addition, the LOD and the specificity were 8.6 pg L^{-1} and 94.44%, respectively. The results were also correlated with those of the gold-standard real-time PCR method, indicating that the fabricated biosensor would be used in clinical applications.

The microcantilever is a mechanical sensor whose material and structure have been greatly improved by developments in the fields of micro- and nanofabrication [187]. Microcantilever sensors have several advantages, including high sensitivity, low-cost, simple operation, and label-free detection, so they can be used as biosensors for DNA hybridization [188], immunoassays [189], and virus detection [190]. For HBV DNA detection, a dynamic microcantilever biosensor enhanced with silica nanoparticles was designed (figure 8.9) [191]. For the DNA target, a 243-mer nucleotide of the HBV DNA precore–core region was used. The detection probe was conjugated with silica nanoparticles, while the capture probe was attached to the microcantilever surface. The detection level of 243-mer HBV target DNA, which was in the picomolar range without the use of nanoparticles, was significantly enhanced by the use of nanoparticles and reached the femtomolar range using the signal amplification strategy. Also, the HBV target DNA concentration provided a linear correlation with the resonant frequency shift. The LOD was 2.3 fM, pointing out that microcantilever biosensors are sensitive and reliable tools for DNA detection.

A Love-wave-mode SAW immunosensor was developed for HBV antibody (HBsAb) detection in whole blood samples. In this design, SiO_2 layers were used as waveguide layers not only to protect the electrode from liquid media, and also to capture acoustic energy near the surface. To achieve a high sensitivity, the guiding layer thickness was thoroughly investigated. The HBV surface antigen (HBsAg) was immobilized on the sensing element to capture HBsAb while the resonance frequency shift was monitored. A reference SAW device was coated with BSA to eliminate the non-specific binding of HBsAb on the surface, minimizing the impact of other potential factors that could affect the mass alterations. The results demonstrated that the sensor was HBsAb specific and there was a linear correlation between the frequency change and the antibody concentration. The biosensor sensitivity was 0.74 Hz (μL^{-1}), and the LOD was less than 10 pg (μL^{-1}) [192]. In a recent study, HBsAg was successfully detected by a PZT integrated assay [193]. Furthermore, various analytes (HBcAg and AFP) were also studied and a dynamic range of measurement concentrations was obtained by the use of multiple reaction sites in the same assay.

8.2.4 Ebola

The ebola virus disease (EVD), originating from a contagious zoonotic infection, was first identified in 1976. The genus *Ebolavirus*, consisting of negative-strand RNA viruses in the *Filoviridae* family, causes hemorrhagic fever that is also transmitted between humans through bodily fluid contact, with an incubation period in the range of 1–21 days [194, 195]. It invades most major organs and causes pathogenic multisystem failure in humans [196]. Five species are known to be responsible for epidemics: Forest ebolavirus, Sudan ebolavirus, Bundibugyo ebolavirus, Reston ebolavirus, and Zaire ebolavirus [195]. EVD is thought to have first caused an infection by direct contact with live or dead animals [197]. The initial symptoms of Ebola are not explicit and they also overlap with other fever-related illnesses; thus, the quick identification of potential cases is crucial to prevent a public health risk holding a mortality rate of 30%–90% [198, 199]. The endemic regions are mainly Central and West Africa, where the availability of medical equipment and trained staff is limited. The current diagnostic methods used for the detection of Ebola viruses could be grouped into three basic categories: (i) serologic tests for the detection of host antibodies IgG [200] or IgM [201], (ii) antigen tests used for the sensing of viral proteins [202] and RT-PCR [203], (iii) molecular (genomic) tests used for the detection of RNA sequences. In particular, sophisticated POC devices could hold the potential to halt future pandemics [204].

From a mechanical biosensor perspective, a reduced graphene oxide (rGO)-based FET platform was used as a biosensor for the real-time detection of the EVD antigen. To improve the semiconducting characteristics of rGO, an AuNP electro-sprayed graphene oxide sheet was used as a conducting channel on the electrodes, and anti-Ebola probes were immobilized to interact with EVD antigens. Part of the rGO sheet passivated by an Al_2O_3 film was created to bridge the gaps between the source and drain electrodes. A fixed drain–source voltage (V_{ds}) was applied between the source and drain electrodes to provide a stable signal. Zaire ebolavirus glycoprotein (EGP) in 0.01 × phosphate-buffered saline (PBS), human serum, and

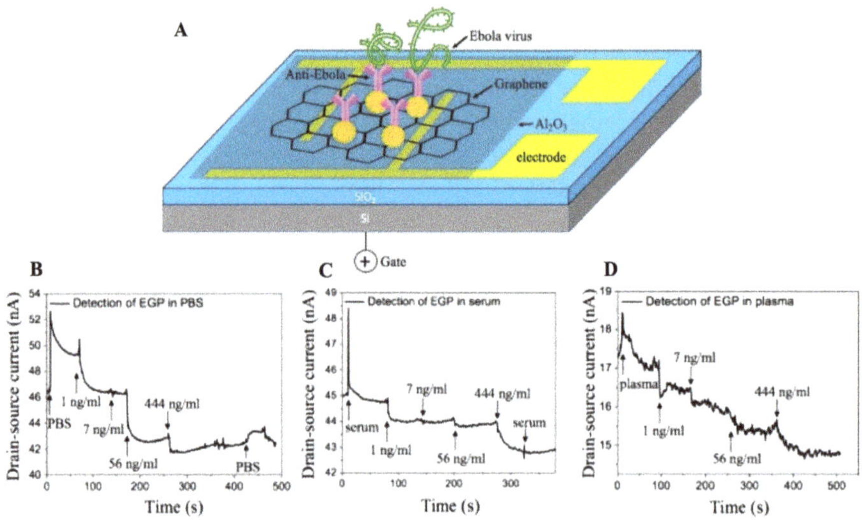

Figure 8.10. (a) A schematic of rGO-based FET biosensor. Dynamic response data of the FET biosensor was obtained from (b) EGP in 0.01 × PBS, (c) EGP in 0.01 x human serum, and (d) EGP in 0.01 × human plasma. Reprinted by permission from Springer (GmbH): Springer Nature, Scientific Reports [205], Copyright (2017).

human plasma solutions were sensed with the LOD of 1 ng mL^{-1}, and the sensitivity increased from 6.5% to 14% (see figure 8.10) [205]. In addition, an SAW sensor and a label-free pathogen-sensing POC device were designed for quick detection of ebola antigens without the necessity for any sample preparation, preparation protocols, or additional reagents. The sensor chips were developed using lithium tantalate ($LiTaO_3$) wafers and a piezoelectric substrate containing IDTs. When an 80–400 MHz alternating voltage was applied to the IDTs, the polarized surface shear waves were propagated horizontally in the piezoelectric substrate. The interactions between the antibody and the viruses emitted an acoustic wave, leading to a frequency alteration that was observed as a log–linear sensor response with the LOD of 1.9×10^4 PFU mL^{-1} [206].

8.2.5 Zika virus

The Zika virus, an arthropod-borne virus (arbovirus), is a member of the *Flaviviridae* family and it is transmitted through infected mosquito bites, similarly to other arboviruses, such as dengue and chikungunya. In addition to direct infection from mosquitoes, infected hosts can also transmit viruses to the fetus during pregnancy, sexual contact, the transfusion of blood and blood products, and organ transplantation [207]. The symptoms are not specific to Zika infection, and this generally leads to misdiagnosis [207]. The common symptoms are known to be a macular or papular rash, fever, arthritis or arthralgia, non-purulent conjunctivitis, myalgia, headache, retro-orbital pain, edema, and vomiting. More rarely, hematospermia, transient dull and metallic hearing, swelling of the hands and ankles, and subcutaneous bleeding may occur [208]. A significant number of patients have mild symptoms, but asymptomatic cases are also known [209].

Diagnosing Zika infection is conventionally carried out using viral nucleic acid (through RT-PCR), IgM antibodies (such as through MAC-ELISA), or a plaque reduction neutralization test (PNT). A definitive diagnosis can be performed using viral nucleic acid from serum; RT-PCR provides conclusive results after one week from the onset of clinical illness [208]. Compared to these methods, a gold-standard diagnosis platform is still a primary necessity. In particular, the massive risk of the cross-reactivity of antibodies for flaviviruses hinders the development of antigenic detection platforms and the interpretation of serologic test results [208, 209]. Moreover, these tests are expensive, hard to conduct, time-consuming, and risky due to the handling of intact viruses. In addition to these limitations, standard reagents are required, which are not easily accessible [208]. Given these facts, the requirement for an accurate, easy-to-use, and cost-efficient detection platform cannot be ignored.

Considering the novel platform requirement, a wide variety of biosensor systems for the diagnosis of Zika infection have been sought by researchers. Micromechanical biosensors are one of the most applied approaches for Zika detection. A recent study, for instance, introduced a QCM biosensor functionalized with aptamers to capture the Zika virus through capsid proteins [210]. The study exhibited three different microbalances, namely a 5 MHz commercial AT-cut (QSense) quartz microbalance, a 10 MHz AT-cut quartz sensor, and a 433 MHz lithium niobate ($LiNbO_3$) sensor. The third microbalance sensor utilized the SAW principle. A comparison was performed between commercially available QCMs and $LiNbO_3$. The sensitivity of the $LiNbO_3$ sensor was determined to be 370 Hz ng^{-1} with the LOD value of 2.7 pg; the sensitivity of the typical commercial 5 MHz QCM was 1 Hz ng^{-1} and its LOD was 1 ng. The introduced system is promising not only for Zika detection, and it would be adapted to detect other virus types [210].

Microfabricated cantilever beams have been employed as biosensors, utilizing the residual stresses upon to mass accumulation on the surface. A recent study demonstrated cantilever beams and aptamers to detect the Zika virus [211]. The surfaces of the microfabricated silicon nitride cantilever beams were coated with gold disks and triangles to mediate the quantification of residual stresses formed when the Zika viruses were bound to the immobilized aptamers. The tensile residual stresses generated were approximately $\sim$ 53 MPa. The stress bent the nitride beams and led to a change in their reflective color and apparent length. The sensitivity of the developed biosensor was observed to be 2 kHz ng^{-1} [211].

8.2.6 Norovirus

The noroviruses are members of the genus *Norovirus* within the family *Caliciviridae*, and they are single-stranded positive-sense RNA viruses. Noroviruses are human enteric pathogens transmitted by food contaminated through feces. Vomiting and person-to-person contact are responsible for at least 50% of all gastroenteritis outbreaks worldwide [212, 213]. Norovirus was first identified in Norwalk, Ohio, in

1968 [214]. The kinetics of norovirus outbreaks seem to be connected with complex interaction variables, including environmental and genetic factors [215]. Due to its enhanced ability to survive in the environment, a small inoculum is required to produce infection (<100 viral particles), for prolonged viral shedding [216]. Electron microscopy was first used for determining viruses, but it is a time-consuming method that cannot be applied to the POC settings [214]. Molecular cloning, enabling capsid protein expression and a comparison of the sequences of caliciviruses has revealed four species: lagovirus, vesivirus, norovirus, and sapovirus [217]. Subsequently, the virus was detected either by electron microscopes or immunologic assay methods, such as ELISA and PCR [218]. These systems lack specificity and sensitivity. Besides, it is critical to detect norovirus quickly due to the shedding rate of 10^5–10^9 noroviruses per gram of feces for 8–60 days [219]. Here, micromechanical biosensor systems could be a breakthrough in addressing such problems mentioned. For instance, a silica microbead-based microfiltration zone and a graphene and gold nanoparticle (Grp-AuNP) composite integrated electrochemical ferrocene tagged aptasensor on a disposable microfluidic platform were developed to detect norovirus [220]. The Grp-AuNP composite provided a durable substrate for aptamer immobilization and microbeads, which filtered out larger particles or solids from the solution. The interactions between the noroviruses and the aptamer led to a decrease in the electrochemical signal originating from the ferrocene. The aptasensor was easily able to distinguish between the target norovirus and other interfering molecules, along with a detection limit of 100 pM and a detection range between 100 pM to 3.5 nM. In another example, a graphene-based FET biosensor composed of a set of antibody-based affinity reagents was used to catch norovirus on a flexible Kapton substrate manufactured by 3D inkjet printing with a sheet resistance as low as 110 Q sq^{-1} [221]. The change in the S12 (i.e. the voltage gain from source to drain) signal intensity at a frequency of 10 GHz provided a linear dose–response relationship from 0.07 to 3.70 dB when the concentration of norovirus VLP protein increased from 0.1 to 100 μg mL^{-1}.

8.2.7 Influenza

Influenza, which belongs to the *Orthomyxoviridae* family, is an acute respiratory disease caused by different influenza viruses. It covers diverse subtypes of the virus, including influenza A, B, and C, differentiated by considering the antigenic differences in their nuclear and matrix proteins [222, 223]. Influenza type C causes relatively milder infections and does not create epidemics, compared to influenza types A and B, which cause respiratory illness epidemics, leading to severe impacts on public health through increased rates of hospitalization and mortality [223].

The genome of the influenza viruses is negative-strand RNA that contains eight genes coding for ten proteins, including the surface proteins hemagglutinin (HA) and neuraminidase (NA) [223]. In addition to these three subtypes, the influenza A virus is categorized according to its hemagglutinin, which is divided into 16 serological subtypes (from H1 to H16), and its neuraminidase (NA), which is divided into 9 serological subtypes (from N1 to N9). Regarding their combinations, approximately 144 possible A/HN subtypes are known; however, not all of them cause diseases in humans. Some of these subtypes have caused pandemic outbreaks, namely A/H1N1 (Spanish flu, 1918), A/H2N2 (Asian flu, 1957), A/H3N2 (Hong Kong flu, 1968), and A/H1N1 again (Swine flu, 1976; Russian flu, 1977) in the last century [224]. In 2009, A/H1N1 caused another pandemic that led to enormous concerns for public health due to the more globalized nature of the world and the pork and poultry production industries [223, 224]. The transmission of these viruses between individuals commonly occurs through large virus-containing droplets and the exchange of aerosols; more rarely, it occurs via contaminated fomites or by direct contact [223].

Several micromechanical platforms have been described in the literature that detect influenza infections aiming to provide low-cost, simplicity, high sensitivity, and rapidity, which would enable the control of transmission and treatment. One of the studies introduced a QCM-based immunosensor to detect influenza types A and B [225]. Here, the QCM sensor employed a frequency or dissipation change due to captured mass. In this manner, antibodies were conjugated with AuNPs to increase the mass sensitivity, and, accordingly, the sensitivity increased tenfold. The LOD value was found to be 1×10^3 PFU mL^{-1}. The sensitivity and specificity of this platform were compared with those of the RT-PCR, shell vial, cell culture, and ELISA methods through a small cohort with 67 clinical samples [225]. Another study also utilized a QCM immunosensor to detect the H5N1 virus, holding an enormous pandemic risk [226]. The study revealed the development of a QCM immunosensor with magnetic nanobeads to enhance the binding reaction between bioreceptor and analytes. Anti-H5N1 polyclonal antibodies were immobilized on a gold-coated QCM surface, which was functionalized with SAMs of 16-mercaptohexadecanoic acid (MHDA) (figure 8.11(a)). When the antibodies captured viruses, a frequency change associated with the concentration occurred (figure 8.11(b)). The developed platform was able to detect the H5N1 virus at a titer of more than 0.0128 HA units within 2 h. The specificity and selectivity of this platform were controlled in the presence of avian influenza subtypes, such as H3N2, H2N2, and H4N8. The sample source was chicken trachea. Consequently, the QCM immunosensor demonstrated huge potential for avian influenza (AI) virus H5N1 detection in clinical, agricultural, environmental, and food samples [226].

Another QCM biosensor was developed to detect avian influenza virus (AIV) H5N1 by combining ssDNA with crosslinked polymeric hydrogel [227]. For the manner of biorecognition, aptamers with high affinity and specificity for AIV H5N1 surface proteins were selected and they formed a hydrogel with the ssDNA. The sensor surface was functionalized with SAMs; the aptamer hydrogel was then immobilized. When the virus was captured by the aptamer, the hydrogel swelled immediately due to the breaking of the bond between the aptamer and the ssDNA; the swelling mediated a frequency shift. The LOD value was determined to be 0.0128 HAUs (HA units) when the aptamer:hydrogel ratio was 1:1. The time required for detection was 30 min, including sampling. This platform offered rapid, sensitive, and specific detection for H5N1 viruses [227].

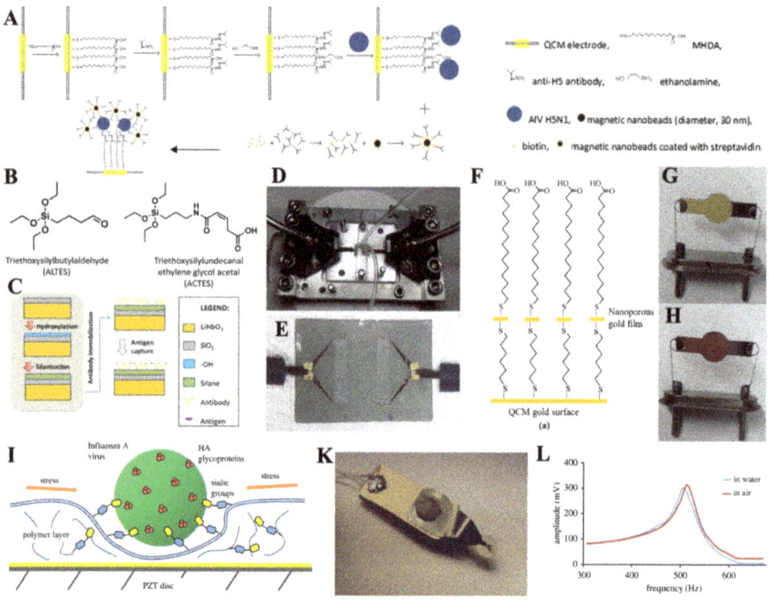

Figure 8.11. (a) A schematic representation of surface functionalization, detection, and amplification. Reprinted from [226], Copyright 2011, with permission from Elsevier. (b) The chemical structure of triethoxysilylbutyraldehyde (ALTES) and triethoxysilylundecanal ethylene glycol acetal (ACTES). (c) The sensor was functionalized via hydroxylation and silanization using n-hydroxysuccinimide (NHS) and ACTES; antibodies were then immobilized. (d), (e) The photograph of an SAW sensor. Reprinted from [228], Copyright 2015, with permission from Elsevier. (f) Nanoporous gold film was functionalized on a gold electrode surface to form a nanowell-based electrode. (g), (h) The photographs of a QCM electrode before (g) and after (h) nanoporous gold film was modified. Reprinted from [52], Copyright 2017, with permission from Elsevier. (i) Lateral stress inside a polymer receptor layer with sialyl oligosaccharide groups increased after the binding of virus particles. (k) A photograph of disc and disc holder. (l) The oscillation data of the piezo disc both in the air (red) and water (blue). Reproduced from [229]. CC BY 4.0.

SAW sensors are another class of micromechanical sensors. A recent study introduced an SAW biosensor, which is capable of detecting the influenza A virus using a Love wave modality [228]. The sensor was fabricated using SiO_2-coated piezoelectric $LiNbO_3$ wafers. The sensor surface was functionalized with triethoxysilylbutylaldehyde (ALTES) and triethoxysilylundecanal ethylene glycol acetal (ACTES), and influenza A HA antibodies were immobilized on the $SiO_2/LiNbO_3$ SAW surface. The LOD value was observed as 1 ng mL^{-1} on the surfaces functionalized with ALTES [228].

Another QCM-based biosensor was introduced to detect H5N1 AIV. In contrast to the other developed platforms, this biosensor was developed using a nanowell (figure 8.11(g) and (h)) structure and an aptamer [52]. The nanowell production employed the metallic corrosion method on a gold film with a ~20 nm pore size. The nanowell electrode was obtained by nanoporous gold film immobilization with a self-assembled monolayer. The ssDNA aptamers were conjugated with NH_2 in 5-terminal and immobilized on the surface via covalent bonding to construct the

QCM aptasensor. When the virus was introduced to the sensor, AIV H5N1 was bound to the immobilized aptamers; therefore, a resonant frequency changed proportional to the virus concentrations. This platform offers label-free detection of H5N1 infection within 10 min. The detection range of the sensor was determined to be 2^{-4} to 2^{4} HAUs 50 μL^{-1} and the LOD value was 2^{-4} HAUs 50 μL^{-1}. The specificity was controlled by applying other subtypes of AIV (e.g., H1N1, H2N2, H7N2, and H5N3). The study also included validation of the sensor using chicken tracheal swab samples. This platform could also be modified and adapted for diverse types of infectious agents [52].

The most recent study introduced a rapid and label-free piezoelectric biosensor to detect influenza A virus by utilizing a PZT piezoelectric disc's first radial mode oscillations. Briefly, PZT was fabricated from a piezoelectric membrane with a radius of 2 mm and a thickness of 100 μm (figure 8.11(k)) [229]. The sensor surface was functionalized with a synthetic sialyl glycopolymer receptor layer, which has an affinity for hemagglutinin proteins (figure 8.11(l)). When viruses were bound to the polymer layer, the disc's radial oscillation mode frequency was increased. The LOD value was determined to be 10^{5} virus particles per milliliter. This platform offers a long lifetime, high sensitivity, selectivity, and easy applicability for home settings [229].

8.3 Conclusions and future perspectives

Viral diseases cause severe health problems for people all over the world. As a result, there is a strong demand for highly sensitive and reliable virus detection methods. Virus detection methods, including immunoassay, molecular detection, and biosensors, have been broadly implemented at multiple settings. A wide range of immunoassay methods, such as ELISA, RIA, ICA, immuno-chemiluminescence, and LFA have been utilized for virus detection. Although they have demonstrated several benefits, their limitations, such as their high-cost and time-consuming nature, have hindered their clinical utility in the early diagnosis of diseases. Molecular detection techniques, such as real-time PCR, nucleic acid tests, and reverse transcription-PCR have many advantages over immunoassay approaches, including higher sensitivity, specificity, and a wider dynamic range. However, these approaches necessitate the use of complicated instruments and pre-processing protocols. Biosensors, on the other hand, are appealing devices for early-stage virus detection, which is a critical step in controlling virus spread and preventing disease. Biosensors offer high sensitivity and specificity, low-cost, real-time analysis, and label-free detection, making them suitable for clinical applications. In particular, this chapter thoroughly investigated various micromechanical sensors, including piezoelectric, surface acoustic wave, and microcantilever biosensors with low detection limits and rapid responses. Furthermore, the most recent developments in micromechanical biosensors for virus detection were extensively reviewed.

Currently, conventional virus detection systems are expensive and only available in centralized clinics and hospitals. Therefore, there is a need to develop a portable POC diagnostic system that can be used by individuals without intensive training. Considering their reliability and rapid response, POC devices are emerging tools for early-stage viral disease diagnosis. Due to micro- and nanotechnological advancements, POC micromechanical biosensors could be an appropriate candidate for virus detection. Micromechanical biosensors conjugated with surface functionalized nanomaterials can be utilized to improve signal amplification, specificity, bioreceptor immobilization, and sensitivity. The signals could be recorded by an individual's smartphone or any wireless devices that transmit the data to clinical health centers and hospitals, so that trained professionals could monitor individuals' health status remotely. Despite the advances in technology and science in the 21st century, the COVID-19 pandemic has demonstrated that virus infection can lead to high morbidity and mortality, hence, a systematic approach to preventing novel pandemic emergence is required urgently. In this respect, artificial intelligence and machine learning are invaluable tools, not only for improving sensor reliability and the reproducibility of sensors, but also for gathering a large volume of information from biosensor responses and analyzing it to create epidemiological models to predict potential pandemics and disease.

In future studies, researchers can focus on developing user-friendly and low-cost biosensors that individuals can perform the tests at home to detect different viruses with a single integrated platform. Another crucial parameter that can be optimized is the detection time, as a person infected with certain viruses, such as COVID-19, can take two to seven days to display symptoms, allowing the virus to be easily transmitted from a patient to a healthy person. Future studies can also concentrate on easy manufacturing label-free biosensors to eliminate the need for harsh reagent storage, or they could develop novel reagents and integrate them into the biosensor structure.

References

[1] Binnig G, Quate C F and Gerber C 1986 Atomic force microscope *Phys. Rev. Lett.* **56** 930

[2] Datar R *et al* 2009 Cantilever sensors: nanomechanical tools for diagnostics *MRS Bull.* **34** 449–54

[3] Ilic B, Yang Y and Craighead H G 2004 Virus detection using nanoelectromechanical devices *Appl. Phys. Lett.* **85** 2604–6

[4] Craighead H G 2000 Nanoelectromechanical systems *Science* **290** 1532–5

[5] Basu A K, Basu A and Bhattacharya S 2020 Micro/nano fabricated cantilever based biosensor platform: a review and recent progress *Enzyme Microb. Technol.* **139** 109558

[6] Kim T S, Yoon D S and Lee J H 2006 Nanomechanical cantilever devices for biological sensors *Micromanufacturing and Nanotechnology* (Berlin: Springer)

[7] Squires T M, Messinger R J and Manalis S 2008 Making it stick: convection, reaction and diffusion in surface-based biosensors *Nat. Biotechnol.* **26** 417–26

[8] Kirstein K U *et al* 2005 Cantilever-based biosensors in CMOS technology *Proc. Design, Automation and Test in Europe* DATE '05 pp 1340–1

[9] Ziegler C 2004 Cantilever-based biosensors *Anal. Bioanal.Chem.* **379** 946–59

[10] Yao D, Fu J and Li F 2021 Measurement of quasi-static and dynamic mechanical properties of soft materials based on instrumented indentation using a piezoelectric cantilever *Polym. Test.* **96** 107056

[11] Thuau D, Ducrot P-H, Poulin P, Dufour I and Ayela C 2018 Integrated electromechanical transduction schemes for polymer MEMS sensors *Micromachines.* **9** 197
[12] Yang H and Gijs M A M 2018 Micro-optics for microfluidic analytical applications *Chem. Soc. Rev.* **47** 1391–458
[13] Enning R, Ziegler D, Nievergelt A, Friedlos R, Venkataramani K and Stemmer A 2011 A high frequency sensor for optical beam deflection atomic force microscopy *Rev. Sci. Instrum.* **82** 43705
[14] Pfreundschuh M, Martinez-Martin D, Mulvihill E, Wegmann S and Muller D J 2014 Multiparametric high-resolution imaging of native proteins by force-distance curve–based AFM *Nat. Protoc.* **9** 1113–30
[15] Johnson B N and Mutharasan R 2012 Sample preparation-free, real-time detection of microRNA in human serum using piezoelectric cantilever biosensors at attomole level *Anal. Chem.* **84** 10426–36
[16] Mishra R and Hegner M 2014 Effect of non-specific species competition from total RNA on the static mode hybridization response of nanomechanical assays of oligonucleotides *Nanotechnology* **25** 225501
[17] Basu A K, Sah A N, Pradhan A and Bhattacharya S 2019 Poly-L-Lysine functionalised MWCNT-rGO nanosheets based 3-d hybrid structure for femtomolar level cholesterol detection using cantilever based sensing platform *Sci. Rep.* **9** 1–13
[18] Bouchaala A, Nayfeh A H and Younis M I 2016 Frequency shifts of micro and nano cantilever beam resonators due to added masses *J. Dyn. Syst. Meas. Control Trans. ASME.* **138**
[19] Martinazzo J, Brezolin A and Steffens C 2020 21st century nanoscience—a handbook: nanopharmaceuticals, nanomedicine, and food nanoscience *Detection of Pesticides Using Cantilever Nanobiosensors* (Boca Raton, FL: CRC Press)
[20] Xie D, Chen D, Peng S, Yang Y, Xu L and Wu F 2019 A low power cantilever-based metal oxide semiconductor gas sensor *IEEE Electron Device Lett.* **40** 1178–81
[21] Agarwal D K, Maheshwari N, Mukherji S and Rao V R 2016 Asymmetric immobilization of antibodies on a piezo-resistive micro-cantilever surface *RSC Adv.* **6** 17606–16
[22] Bai X, Hou H, Zhang B and Tang J 2014 Label-free detection of kanamycin using aptamer-based cantilever array sensor *Biosens. Bioelectron.* **56** 112–6
[23] Chen Y, Xu P and Li X 2010 Self-assembling siloxane bilayer directly on SiO_2 surface of micro-cantilevers for long-term highly repeatable sensing to trace explosives *Nanotechnology* **21** 265501
[24] Andrade M A *et al* 2020 A nanomechanical genosensor using functionalized cantilevers to detect the cancer biomarkers miRNA-203 and miRNA-205 *IEEE Sens. J.* **20** 2860–7
[25] Nyang'au W O, Setiono A, Schmidt A, Bosse H and Peiner E 2020 Sampling and mass detection of a countable number of microparticles using on-cantilever imprinting *Sensors* **20** 2508
[26] Gad-el-Hak M 2001 *The MEMS Handbook* (Boca Raton, FL: CRC Press)
[27] Beeby S, Ensel G, White N and Kraft M 2004 *MEMS Mechanical Sensors* (Boston, MA: Artech House)
[28] Riesch C, Keplinger F, Reichel E K and Jakoby B 2006 Characterizing resonating cantilevers for liquid property sensing *Proc. IEE Sensors* pp 1070–3
[29] Brand O, Dufour I, Heinrich S M and Josse F 2015 Resonant MEMS: principles, modeling, implementation, and applications *Resonant MEMS: Principles, Modeling, Implementation, and Applications.* (New York: Wiley) pp 1–483
[30] Sader J E 1998 Frequency response of cantilever beams immersed in viscous fluids with applications to the atomic force microscope *J. Appl. Phys.* **84** 64–76

[31] Van Eysden C A and Sader J E 2009 Frequency response of cantilever beams immersed in compressible fluids with applications to the atomic force microscope *J. Appl. Phys.* **106** 094904

[32] Raiteri R, Grattarola M, Butt H J and Skládal P 2001 Micromechanical cantilever-based biosensors *Sensors Actuators* B **79** 115–26

[33] Shiraishi N, Ikehara T, Dao D V, Sugiyama S and Ando Y 2013 Fabrication and testing of polymer cantilevers for VOC sensors *Sensors Actuators* A **02** 233–9

[34] Abadal G *et al* 2001 Electromechanical model of a resonating nano-cantilever-based sensor for high-resolution and high-sensitivity mass detection *Nanotechnology* **12** 100

[35] Durdaut P *et al* 2020 Fundamental noise limits and sensitivity of piezoelectrically driven magnetoelastic cantilevers *J. Microelectromech. Syst.* **29** 1347–61

[36] Teva J *et al* 2004 On the electromechanical modelling of a resonating nano-cantilever-based transducer *Ultramicroscopy* (Amsterdam: North-Holland) pp 225–32

[37] Park J, Nishida S, Lambert P, Kawakatsu H and Fujita H 2011 High-resolution cantilever biosensor resonating at air-liquid in a microchannel *Lab Chip.* **11** 4187–93

[38] Yang Y-T, Callegari C, Feng X L, Ekinci K L and Roukes M L 2006 Zeptogram-scale nanomechanical mass sensing *Nano Lett.* **6** 583–6

[39] Ilic B, Yang Y, Aubin K, Reichenbach R, Krylov S and Craighead H G 2005 Enumeration of DNA molecules bound to a nanomechanical oscillator *Nano Lett.* **5** 925–9

[40] Fu Y Q *et al* 2017 Advances in piezoelectric thin films for acoustic biosensors, acousto-fluidics and lab-on-chip applications *Prog. Mater Sci.* **89** 31–91

[41] Sauerbrey G 1959 Use of quartz vibrator for weighting thin films on a microbalance *Z. Phys.* **155** 206–12

[42] Zhang J, Zhang X, Wei X, Xue Y, Wan H and Wang P 2021 Recent advances in acoustic wave biosensors for the detection of disease-related biomarkers: a review *Anal Chim Acta* **1164** 338321

[43] Murillo A E *et al* 2019 Development of AlN thin films for breast cancer acoustic biosensors *J. Mater. Res. Technol.* **8** 350–8

[44] Fogel R *et al* 2019 Bulk and surface acoustic wave sensor arrays for multi-analyte detection: a review *Biosens. Bioelectron.* **163** 112164

[45] Zhang Y, Luo J, Flewitt A J, Cai Z and Zhao X 2018 Film bulk acoustic resonators (FBARs) as biosensors: a review *Biosens. Bioelectron.* **116** 1–15

[46] Afzal A, Mujahid A, Schirhagl R, Bajwa S Z, Latif U and Feroz S 2017 Gravimetric viral diagnostics: QCM based biosensors for early detection of viruses *Chemosensors.* **5** 7

[47] Dey N, Ashour A S, Mohamed W S and Nguyen N G 2019 Acoustic sensors *Acoustic Sensors for Biomedical Applications* (Berlin: Springer) pp 33–41

[48] Salouti M and Derakhshan F K 2020 Biosensors and nanobiosensors in environmental applications *Biogenic Nano-Particles and their Use in Agro-Ecosystems* (Berlin: Springer) pp 515–91

[49] Pirinçci Ş, Ertekin Ö, Laguna D E, Özen F, Öztürk Z Z and Öztürk S 2018 Label-free QCM immunosensor for the detection of ochratoxin A *Sensors* **18** 1161

[50] Yu X, Chen F, Wang R and Li Y 2018 Whole-bacterium SELEX of DNA aptamers for rapid detection of *E.coli* O157:H7 using a QCM sensor *J. Biotechnol.* **266** 39–49

[51] Yilmaz E, Majidi D, Ozgur E and Denizli A 2015 Whole cell imprinting based *Escherichia coli* sensors: a study for SPR and QCM *Sensors Actuators* B **209** 714–21

[52] Wang R, Wang L, Callaway Z T, Lu H, Huang T J and Li Y 2017 A nanowell-based QCM aptasensor for rapid and sensitive detection of avian influenza virus *Sensors Actuators* B **240** 934–40

[53] Wang R and Li Y 2013 Hydrogel based QCM aptasensor for detection of avian influenzavirus *Biosens. Bioelectron.* **42** 148–55

[54] Arif S, Qudsia S, Urooj S, Chaudry N, Arshad A and Andleeb S 2015 Blueprint of quartz crystal microbalance biosensor for early detection of breast cancer through salivary autoantibodies against ATP6AP1 *Biosens. Bioelectron.* **65** 62–70

[55] Liu J *et al* 2020 A microfabricated thickness shear mode electroacoustic resonator for the label-free detection of cardiac troponin in serum *Talanta* **215** 120890

[56] Mujahid A, Afzal A and Dickert F L 2019 An overview of high frequency acoustic sensors —QCMs, SAWs and FBARs—chemical and biochemical applications *Sensors* **19** 4395

[57] Ren J, Chu H, Bai Y, Wang R, Chen P and Chen J 2021 Research and design of high sensitivity FBAR micro-mass sensors *IOP Conf. Series: Earth and Environmental Science (Bristol) (IOP Publishing)* pp 42014

[58] Liu C, Pang W, Duan X and Wang Y 2020 Dual functions of Ghz frequency acoustic resonator system for biosamples capture and sensing *2020 42nd Annual Int. Conf. of the IEEE Engineering in Medicine & Biology Society (EMBC) (Piscataway, NJ)* (IEEE) pp 3994–7

[59] White R M and Voltmer F W 1965 Direct piezoelectric coupling to surface elastic waves *Appl. Phys. Lett.* **7** 314–6

[60] Xu Z and Yuan Y J 2018 Implementation of guiding layers of surface acoustic wave devices: a review *Biosens. Bioelectron.* **99** 500–12

[61] Raj V B, Singh H, Nimal A T, Sharma M U and Gupta V 2013 Oxide thin films (ZnO, TeO_2, SnO_2, and TiO_2) based surface acoustic wave (SAW) E-nose for the detection of chemical warfare agents *Sensors Actuators* B **178** 636–47

[62] Constantinoiu I and Viespe C 2019 Detection of volatile organic compounds using surface acoustic wave sensor based on nanoparticles incorporated in polymer *Coatings.* **9** 373

[63] Chen C and Jin J 2020 Surface acoustic wave vapor sensor with graphene interdigital transducer for TNT detection *Sens. Imag.* **21** 1–12

[64] Al-Mashat L, Tran H D, Wlodarski W, Kaner R B and Kalantar-Zadeh K 2008 Polypyrrole nanofiber surface acoustic wave gas sensors *Sensors Actuators* B **134** 826–31

[65] Kidakova A, Boroznjak R, Reut J, Öpik A, Saarma M and Syritski V 2020 Molecularly imprinted polymer-based SAW sensor for label-free detection of cerebral dopamine neurotrophic factor protein *Sensors Actuators* B **308** 127708

[66] Wang K *et al* 2018 Sorting of tumour cells in a microfluidic device by multi-stage surface acoustic waves *Sensors Actuators* B **258** 1174–83

[67] Go D B, Atashbar M Z, Ramshani Z and Chang H-C 2017 Surface acoustic wave devices for chemical sensing and microfluidics: a review and perspective *Anal. Methods* **9** 4112–34

[68] Lamanna L, Rizzi F, Bhethanabotla V R and De Vittorio M 2020 Conformable surface acoustic wave biosensor for *E.coli* fabricated on PEN plastic film *Biosens. Bioelectron.* **163** 112164

[69] Bisoffi M *et al* 2008 Detection of viral bioagents using a shear horizontal surface acoustic wave biosensor *Biosens. Bioelectron.* **23** 1397–403

[70] Luo J, Quan A, Fu C and Li H 2017 Shear-horizontal surface acoustic wave characteristics of a (110) ZnO/SiO2/Si multilayer structure *J. Alloys Compd.* **693** 558–64

[71] Länge K, Rapp B E and Rapp M 2008 Surface acoustic wave biosensors: a review *Anal. Bioanal. Chem.* **391** 1509–19

[72] Puiu M, Gurban A-M, Rotariu L, Brajnicov S, Viespe C and Bala C 2015 Enhanced sensitive Love wave surface acoustic wave sensor designed for immunoassay formats *Sensors* **15** 10511–25

[73] Devkota J, Ohodnicki P R and Greve D W 2017 SAW sensors for chemical vapors and gases *Sensors* **17** 801
[74] Mohseni H and Ng C-T 2019 Rayleigh wave propagation and scattering characteristics at debondings in fibre-reinforced polymer-retrofitted concrete structures *Struct. Heal. Monit.* **18** 303–17
[75] Agostini M, Greco G and Cecchini M 2018 A Rayleigh surface acoustic wave (R-SAW) resonator biosensor based on positive and negative reflectors with sub-nanomolar limit of detection *Sensors Actuators* B **254** 1–7
[76] Pantazis A K, Konstantinidis G and Gizeli E 2016 Study of the effect of the operating frequency of a GaN Lamb wave device to viscosity and protein sensing *IEEE Sens. J.* **16** 7028–36
[77] Nam J, Jang W S, Kim J, Lee H and Lim C S 2019 Lamb wave-based molecular diagnosis using DNA hydrogel formation by rolling circle amplification (RCA) process *Biosens. Bioelectron* **142** 111496
[78] Hamidullah M, Élie-Caille C and Leblois T 2020 Theoretical study and finite element simulation of ZnO/GaAs higher-order Lamb waves for microsensor application in liquid media *IEEE Sensors* (Piscataway, NJ: IEEE) pp 1–4
[79] Uchino K and Debus J C 2012 *Applications of ATILA FEM Software to Smart Materials: Case Studies in Designing Devices.* (Cambridge: Woodhead) pp 1–386
[80] Curry E J *et al* 2018 Biodegradable piezoelectric force sensor *Proc. Natl Acad. Sci.* **115** 909–14
[81] Lee M *et al* 2012 A hybrid piezoelectric structure for wearable nanogenerators *Adv. Mater.* **24** 1759–64
[82] Su L, Fong C C, Cheung P Y and Yang M 2017 Development of novel piezoelectric biosensor using PZT ceramic resonator for detection of cancer markers *Methods in Molecular Biology* (Totowa, NJ: Humana Press) pp 277–91
[83] Zhu B *et al* 2017 Self-focused AlScN film ultrasound transducer for individual cell manipulation *ACS Sens.* **2** 172–7
[84] Rajala S *et al* 2016 Cellulose nanofibril film as a piezoelectric sensor material *ACS Appl. Mater. Interfaces* **8** 15607–14
[85] Ghosh S K and Mandal D 2017 Sustainable energy generation from piezoelectric biomaterial for noninvasive physiological signal monitoring *ACS Sustain Chem Eng.* **5** 8836–43
[86] Ghosh S K and Mandal D 2017 Bio-assembled, piezoelectric prawn shell made self-powered wearable sensor for non-invasive physiological signal monitoring *Appl. Phys. Lett.* **110** 123701
[87] Pohanka M 2018 Piezoelectric biosensor for the determination of tumor necrosis factor alpha *Talanta* **178** 970–3
[88] Chorsi M T *et al* 2019 Piezoelectric biomaterials for sensors and actuators *Adv. Mater.* **31** 1–15
[89] Janshoff A, Galla H J and Steinem C 2000 Piezoelectric mass-sensing devices as biosensors – an alternative to optical biosensors? *Angew Chem. – Int. Ed.* **39** 4004–32
[90] Ramadan K S, Sameoto D and Evoy S 2014 A review of piezoelectric polymers as functional materials for electromechanical transducers *Smart Mater. Struct.* **23** 033001
[91] Algamili A S *et al* 2021 A review of actuation and sensing mechanisms in MEMS-based sensor devices *Nanoscale Res. Lett.* **16** 16
[92] Pang W, Zhao H, Kim E S, Zhang H, Yu H and Hu X 2012 Piezoelectric micro-electromechanical resonant sensors for chemical and biological detection *Lab Chip* **12** 29–44

[93] Huang I, Lee M, Hsu C and Wang C 2012 Chemical development of a flexural plate-wave (FPW) immunoglobulin-E (IgE) allergy bio-sensing microsystem *Sensors Actuators* B **162** 184–93
[94] Jung J, Lee W, Kang W, Shin E, Ryu J and Choi H 2017 Review of piezoelectric micromachined ultrasonic transducers and their applications *J Micromech. Microeng.* **27** 113001
[95] Pohanka M 2018 Overview of piezoelectric biosensors, immunosensors and DNA sensors and their applications *Materials (Basel)* **11** 448
[96] Zhang J X J and Hoshino K 2019 Mechanical transducers: cantilevers, acoustic wave sensors, and thermal sensors *Molecular Sensors and Nanodevices* (Amsterdam: Elsevier) pp 311–412
[97] Fritz J *et al* 2000 Translating biomolecular recognition into nanomechanics *Science* **288** 316–8
[98] Gibbs J W and Josiah W 1839–903 *Bumstead HA, 1870-920, Van Name RG, 1877–. Scientific Papers of J. Willard Gibbs* (London: Longmans, Green and Co.) 1906
[99] Thundat T, Oden P I and Warmack R J 1997 Microcantilever sensors *Microscale Thermophys. Eng.* **1** 185–99
[100] Butt H J 1996 A sensitive method to measure changes in the surface stress of solids *J. Colloid Interface Sci.* **180** 251–60
[101] Mathew R, Ravi and Sankar A 2018 A review on surface stress-based miniaturized piezoresistive SU-8 polymeric cantilever sensors *Nano-Micro Letters* vol 10 (Berlin: Springer) pp 1–41
[102] Stoney G G 1909 The tension of metallic films deposited by electrolysis *Proc. R Soc. London Ser.* A **82** 172–5
[103] Liangruksa M 2008 *Effect of Surface Stress on Micromechanical Cantilevers for Sensing Applications.* (Blacksburg, VA: Virginia Tech)
[104] Sang S, Zhao Y, Zhang W, Li P, Hu J and Li G 2014 Surface stress-based biosensors *Biosens. Bioelectron.* **51** 124–35
[105] Lundbæk J A, Maer A M and Andersen O S 1997 Lipid bilayer electrostatic energy, curvature stress, and assembly of gramicidin channels *Biochemistry.* **36** 5695–701
[106] Berger R *et al* 1997 Surface stress in the self-assembly of alkanethiols on gold *Science* **276** 2021–4
[107] Jian A *et al* 2017 A PDMS surface stress biosensor with optimized micro-membrane: fabrication and application *Sensors Actuators* B **242** 969–76
[108] Takahashi K, Oyama H, Misawa N, Okumura K, Ishida M and Sawada K 2013 Surface stress sensor using MEMS-based Fabry–Perot interferometer for label-free biosensing *Sensors Actuators* B **188** 393–9
[109] Yue M *et al* 2004 A 2-D microcantilever array for multiplexed biomolecular analysis *J. Microelectromech. Syst.* **13** 290–9
[110] Marie R, Jensenius H, Thaysen J, Christensen C B and Boisen A 2002 Adsorption kinetics and mechanical properties of thiol-modified DNA-oligos on gold investigated by micro-cantilever sensors *Ultramicroscopy* **91** 29–36
[111] Varshney M, Li Y, Srinivasan B and Tung S 2007 A label-free, microfluidics and interdigitated array microelectrode-based impedance biosensor in combination with nano-particles immunoseparation for detection of *Escherichia coli* O157:H7 in food samples *Sensors Actuators* B **128** 99–107
[112] Choi Y J, Takahashi T, Taki M, Sawada K and Takahashi K 2021 Label-free attomolar protein detection using a MEMS optical interferometric surface-stress immunosensor with a freestanding PMMA/parylene-C nanosheet *Biosens. Bioelectron.* **172** 112778

[113] Mertens J *et al* 2008 Label-free detection of DNA hybridization based on hydration-induced tension in nucleic acid films *Nat. Nanotechnol.* **3** 301–7

[114] Godin M *et al* 2004 Surface stress, kinetics, and structure of alkanethiol self-assembled monolayers *Langmuir* **20** 7090–6

[115] Battiston F M *et al* 2001 A chemical sensor based on a microfabricated cantilever array with simultaneous resonance-frequency and bending readout *Sensors Actuators* B **77** 122–31

[116] Lim Y, Kouzani A and Duan W *et al* 2014 A surface-stress-based microcantilever aptasensor *IEEE Trans. on Biomed. Circ. Sys.* **8** 15–24

[117] Yue M 2004 *Multiplexed Label-Free Bioassays Based on Nanomechanics and Nanofluidics* (Ann Arbor, MI: ProQuest Dissertations Publishing)

[118] Zhang W *et al* 2013 Structural optimization of the micro-membrane for a novel surface stress-based capacitive biosensor *Microelectron. Eng.* **106** 9–12

[119] Benderitter M, Vincent-Genod L, Pouget J P and Voisin P 2003 The cell membrane as a biosensor of oxidative stress induced by radiation exposure: a multiparameter investigation *Radiat. Res.* **159** 471–83

[120] Zhao D *et al* 2019 Sandwich-type surface stress biosensor based on self-assembled gold nanoparticles in PDMS film for BSA detection *ACS Biomater. Sci. Eng.* **5** 6274–80

[121] Sang S and Witte H 2010 A novel PDMS micro membrane biosensor based on the analysis of surface stress *Biosens. Bioelectron.* **25** 2420–4

[122] Yang S M and Chang C 2010 A piezoresistive bridge-microcantilever biosensor by CMOS process for surface stress measurement *Sensors Actuators* B **145** 405–10

[123] Zhao D *et al* 2019 Surface stress-based biosensor with stable conductive AuNPs network for biomolecules detection *Appl. Surf. Sci.* **491** 443–50

[124] Doelle M, Mager D, Ruther P and Paul O 2006 Geometry optimization for planar piezoresistive stress sensors based on the pseudo-Hall effect *Sensors Actuators* A **127** 261–9

[125] Takahashi T, Hizawa T, Misawa N, Taki M, Sawada K and Takahashi K 2018 Surface stress sensor based on MEMS Fabry–Perot interferometer with high wavelength selectivity for label-free biosensing *J. Micromech. Microeng.* **28** 054002

[126] Yoshikawa G, Akiyama T, Gautsch S, Vettiger P and Rohrer H 2011 Nanomechanical membrane-type surface stress sensor *Nano Lett.* **11** 1044–8

[127] Jones K E *et al* 2008 Global trends in emerging infectious diseases *Nature* **451** 990–3

[128] Qureshi A and Niazi J H 2020 Biosensors for detecting viral and bacterial infections using host biomarkers: a review *Analyst* **145** 7825–48

[129] Narita F *et al* 2021 A review of piezoelectric and magnetostrictive biosensor materials for detection of COVID-19 and other viruses *Adv. Mater.* **33** 2005448

[130] Rodovalho V, Alves L, Castro A, Madurro J, Brito-Madurro A and Santos A 2015 Biosensors applied to diagnosis of infectious diseases—an update *Austin J. Biosens. Bioelectron.* **1** 1–12

[131] Sin M L Y, Mach K E, Wong P K and Liao J C 2014 Advances and challenges in biosensor-based diagnosis of infectious diseases *Expert Rev. Mol. Diagn.* **14** 225–44

[132] Song Z *et al* 2019 From SARS to MERS, thrusting coronaviruses into the spotlight *Viruses.* **11** 59

[133] Yadav S *et al* 2021 SERS based lateral flow immunoassay for point-of-care detection of SARS-CoV-2 in clinical samples *ACS Appl Bio Mater.* **4** 2974–95

[134] Zhong N S *et al* 2020 Epidemiology and cause of severe acute respiratory syndrome (SARS) in Guangdong, People's Republic of China, in February, 2003 Elsevier Connect, the company's public news and information *Lancet* **362** 1353–8

[135] Alyami M H, Alyami H S and Warraich A 2020 Middle East Respiratory Syndrome (MERS) and novel coronavirus disease-2019 (COVID-19): from causes to preventions in Saudi Arabia *Saudi Pharm J.* **28** 1481–91

[136] Butowt R and Bilinska K 2020 SARS-CoV-2: olfaction, brain infection, and the urgent need for clinical samples allowing earlier virus detection *ACS Chem. Neurosci.* **11** 1200–3

[137] Chan J F W *et al* 2020 A familial cluster of pneumonia associated with the 2019 novel coronavirus indicating person-to-person transmission: a study of a family cluster *Lancet* **395** 514–23

[138] Li T *et al* 2021 Computer-aided diagnosis of COVID-19 CT scans based on spatiotemporal information fusion *J. Healthc. Eng.* **2021** 6649591

[139] Ruuskanen O, Lahti E, Jennings L C and Murdoch D 2011 Viral pneumonia *Lancet* **377** 1264–75

[140] Cui F and Zhou H S 2020 Diagnostic methods and potential portable biosensors for coronavirus disease 2019 *Biosens. Bioelectron.* **165** 112349

[141] Liu W *et al* 2020 Evaluation of nucleocapsid and spike protein-based enzyme-linked immunosorbent assays for detecting antibodies against SARS-CoV-2 *J. Clin. Microbiol.* **58** 6

[142] Lu R, Wu X, Wan Z, Li Y, Jin X and Zhang C 2020 A novel reverse transcription loop-mediated isothermal amplification method for rapid detection of SARS-CoV-2 *Int. J. Mol. Sci.* **21** 2826

[143] Xia N, Wang G and Gong W 2020 Serological test is an efficient supplement for detecting RNA to confirm SARS-CoV-2 infection *Preprint.* 1–6

[144] Broughton J P *et al* 2020 Rapid detection of 2019 novel coronavirus SARS-CoV-2 using a CRISPR-based DETECTR lateral flow assay *medRxiv* **415** 1–28

[145] Fang Y and Pang P 2020 Senivity of chest CT for COVID-19: comparasion to RT-PCR *Radiology* **296** 15–7

[146] Pan Y *et al* 2020 Serological immunochromatographic approach in diagnosis with SARS-CoV-2 infected COVID-19 patients *J. Infect.* **81** e28–32

[147] Wang Y, Kang H, Liu X and Tong Z 2020 Combination of RT-qPCR testing and clinical features for diagnosis of COVID-19 facilitates management of SARS-CoV-2 outbreak *J. Med. Virol.* **92** 538–9

[148] Ranjan P *et al* 2020 Biosensor-based diagnostic approaches for various cellular biomarkers of breast cancer: a comprehensive review *Anal. Biochem.* **610** 113996

[149] Demeke Teklemariam A, Samaddar M, Alharbi M G, Al-Hindi R R and Bhunia A K 2020 Biosensor and molecular-based methods for the detection of human coronaviruses: a review *Mol. Cell. Probes* **54** 1–8

[150] Zuo B, Li S, Guo Z, Zhang J and Chen C 2004 Piezoelectric immunosensor for SARS-associated coronavirus in sputum *Anal. Chem.* **76** 3536–40

[151] Velanki S and Ji H F 2006 Detection of feline coronavirus using microcantilever sensors *Meas. Sci. Technol.* **17** 2964–8

[152] Du F and Zhang F 2020 Application of novel MEMS sensor in airborne COVID-19 virus detection *Sci. J. Info. Eng.* **9** 1–5

[153] Deeks S G, Overbaugh J, Phillips A and Buchbinder S 2015 HIV infection *Nat. Rev. Dis. Prim.* **1**

[154] World Health Organization HIV/AIDS (https://who.int/news-room/fact-sheets/detail/hiv-aids) (Accessed: on 4th April 2021)

[155] Lam Y, Abu-Lail N I, Alam M S and Zauscher S 2006 Using microcantilever deflection to detect HIV-1 envelope glycoprotein gp120 *Nanomed. Nanotech. Biol Med.* **2** 222–9

[156] Sheikh S, Blaszykowski C and Thompson M 2011 Label-free detection of HIV-2 antibodies in serum with an ultra-high frequency acoustic wave sensor *Talanta* **85** 816–9

[157] Lu C H *et al* 2012 Sensing HIV related protein using epitope imprinted hydrophilic polymer coated quartz crystal microbalance *Biosens. Bioelectron.* **31** 439–44

[158] Ly T N, Park S and Park S J 2016 Detection of HIV-1 antigen by quartz crystal microbalance using gold nanoparticles *Sensors Actuators* B **237** 452–8

[159] Bisoffi M, Severns V, Branch D W, Edwards T L and Larson R S 2013 Rapid detection of human immunodeficiency virus types 1 and 2 by use of an improved piezoelectric biosensor *J. Clin. Microbiol.* **51** 1685–91

[160] Saurabh N K, Deekshanya B and Rudraswamy S B 2017 Design of microcantilever sensor for detection of P24 antigen in HIV diagnoses *Int. J. Sci. Eng. Sci.* 1

[161] Zhou J *et al* 2017 Assessing activity of hepatitis A virus 3C protease using a cyclized luciferase-based biosensor *Biochem. Biophys. Res. Commun.* **488** 621–7

[162] Shepard C W, Simard E P, Finelli L, Fiore A E and Bell B P 2006 Hepatitis B virus infection: epidemiology and vaccination *Epidemiol Rev.* **28** 112–25

[163] Donmez S, Arslan F and Arslan H 2015 A nucleic acid biosensor for detection of hepatitis C virus genotype 1a using poly(L-Glutamic acid)-modified electrode *Appl. Biochem. Biotechnol.* **176** 1431–44

[164] Donmez S, Arslan F and Arslan H 2016 Sequence-specific label-free nucleic acid biosensor for the detection of the hepatitis C virus genotype 1a using a disposable pencil graphite electrode *Artif. Cells, Nanomed. Biotechnol.* **44** 912–7

[165] Shawky S M, Awad A M, Allam W, Alkordi M H and EL-Khamisy S F 2017 Gold aggregating gold: a novel nanoparticle biosensor approach for the direct quantification of hepatitis C virus RNA in clinical samples *Biosens. Bioelectron.* **92** 349–56

[166] Nainan O V, Xia G, Vaughan G and Margolis H S 2006 Diagnosis of hepatitis A virus infection: a molecular approach *Clin Microbiol Rev.* **19** 63–79

[167] Wang X *et al* 2010 Gold nanorod-based localized surface plasmon resonance biosensor for sensitive detection of hepatitis B virus in buffer, blood serum and plasma *Biosens. Bioelectron.* **26** 404–10

[168] Mathiesen L R, Feinstone S M, Wong D C, Skinhoej P and Purcell R H 1978 Enzyme-linked immunosorbent assay for detection of hepatitis A antigen in stool and antibody to hepatitis A antigen in sera: comparison with solid-phase radioimmunoassay, immune electron microscopy, and immune adherence hemagglutination assay *J. Clin. Microbiol.* **7** 184–93

[169] Gocke D J and Kachani Z F 1973 Solid-phase radioimmunoassay for hepatitis B antigen *JAMA J. Am. Med. Assoc.* **224** 1425–6

[170] Shen J *et al* 2015 Immunochromatographic assay for quantitative and sensitive detection of hepatitis B virus surface antigen using highly luminescent quantum dot-beads *Talanta* **142** 145–9

[171] Sabouri S, Ghourchian H, Shourian M and Boutorabi M 2014 A gold nanoparticle-based immunosensor for the chemiluminescence detection of the hepatitis B surface antigen *Anal. Methods* **6** 5059–66

[172] Ryu J H *et al* 2018 Development of a rapid automated fluorescent lateral flow immunoassay to detect hepatitis B surface antigen (HBsAg), antibody to HBsAg, and antibody to hepatitis C *Ann Lab Med.* **38** 578

[173] Peruski A H and Peruski L F 2003 Immunological methods for detection and identification of infectious disease and biological warfare agents *Clin Diagn Lab Immunol.* **10** 506–13

[174] Zhang C *et al* 2021 Rapid one-pot detection of SARS-CoV-2 based on a lateral flow assay in clinical samples *Anal. Chem.* **93** 3325–30

[175] Cobo F 2012 Application of molecular diagnostic techniques for viral testing *Open Virol J.* **5** 104–14

[176] Khodakov D A, Zakharova N V, Gryadunov D A, Filatov F P, Zasedatelev A S and Mikhailovich V M 2008 An oligonucleotide microarray for multiplex real-time PCR identification of HIV-1, HBV, and HCV *Biotechniques.* **44** 241–8

[177] de Vera M E *et al* 2018 Transplantation of hepatitis C virus (HCV) antibody positive, nucleic acid test negative donor kidneys to HCV negative patients frequently results in seroconversion but not HCV viremia *Am. J. Transplant.* **18** 2451–6

[178] Narayanamurthy V, Jeroish Z E, Bhuvaneshwari K S and Samsuri F 2021 Hepatitis C virus (HCV) diagnosis: via microfluidics *Anal. Methods* **13** 740–63

[179] Parizad E and Khosravi A 2012 Detection of HBV DNA in cerumen and sera of Hbsag negative patients with chronic hepatitis B infection *Iran Red Crescent* **14** 186–7

[180] Srisomwat C, Teengam P, Chuaypen N, Tangkijvanich P, Vilaivan T and Chailapakul O 2020 Pop-up paper electrochemical device for label-free hepatitis B virus DNA detection *Sensors Actuators* B **316** 128077

[181] Yoo H, Shin J, Sim J, Cho H and Hong S 2020 Biosensors and bioelectronics reusable surface plasmon resonance biosensor chip for the detection of H1N1 influenza virus *Biosens. Bioelectron.* **168** 112561

[182] Perumal V and Hashim U 2014 Advances in biosensors: principle, architecture and applications *J. Appl. Biomed.* **12** 1–15

[183] Liu Y P and Yao C Y 2015 Rapid and quantitative detection of hepatitis B virus *World J Gastroenterol.* **21** 11954–63

[184] Lim H J, Saha T, Tey B T, Tan W S and Ooi C W 2020 Quartz crystal microbalance-based biosensors as rapid diagnostic devices for infectious diseases *Biosens. Bioelectron.* **168** 112513

[185] Zhou X, Liu L, Hu M, Wang L and Hu J 2002 Detection of hepatitis B virus by piezoelectric biosensor *J. Pharm. Biomed. Anal.* **27** 341–5

[186] Yao C *et al* 2008 Hybridization assay of hepatitis B virus by QCM peptide nucleic acid biosensor *Biosens. Bioelectron.* **23** 879–85

[187] Zhao J *et al* 2021 Gold nanoparticles amplified microcantilever biosensor for detecting protein biomarkers with high sensitivity *Sensors Actuators* A **321** 112563

[188] Wu G 2001 Origin of nanomechanical cantilever motion generated from biomolecular interactions *Proc. Natl Acad. Sci.* **98** 1560–4

[189] Wu G, Datar R H, Hansen K M, Thundat T, Cote R J and Majumdar A 2001 Bioassay of prostate-specific antigen (PSA) using microcantilevers *Nat. Biotechnol.* **19** 856–60

[190] Timurdogan E, Alaca B E, Kavakli I H and Urey H 2011 MEMS biosensor for detection of hepatitis A and C viruses in serum *Biosens. Bioelectron.* **28** 189–94

[191] Cha B H *et al* 2009 Detection of hepatitis B virus (HBV) DNA at femtomolar concentrations using a silica nanoparticle-enhanced microcantilever sensor *Biosens. Bioelectron.* **25** 130–5

[192] Lee H J, Namkoong K, Cho E C, Ko C, Park J C and Lee S S 2009 Surface acoustic wave immunosensor for real-time detection of hepatitis B surface antibodies in whole blood samples *Biosens. Bioelectron.* **24** 3120–5

[193] Xu T, Miao J, Wang Z, Yu L and Li C M 2011 Micro-piezoelectric immunoassay chip for simultaneous detection of hepatitis B virus and α-fetoprotein *Sensors Actuators* B **151** 370–6

[194] WHO Ebola Response Team 2014 Ebola virus disease in West Africa—The first 9 months of the epidemic and forward projections *N. Engl. J. Med.* **371** 1481–95

[195] Beeching N J, Fenech M and Houlihan C F 2014 Ebola virus disease *BMJ (Online)* **349** g7348

[196] Mahanty S and Bray M 2004 Pathogenesis of filoviral haemorrhagic fevers *The Lancet Infec. Dis.* **4** 487–98

[197] Pourrut X *et al* 2005 The natural history of Ebola virus in Africa *Microbes Infect.* **7** 1005–14

[198] WHO 2021 Ebola Virus Disease. (https://www.who.int/news-room/fact-sheets/detail/ebola-virus-disease)

[199] Chertow D S, Kleine C, Edwards J K, Scaini R, Giuliani R and Sprecher A 2014 Ebola virus disease in West Africa – clinical manifestations and management *N. Engl. J. Med.* **371** 2054–7

[200] Brangel P *et al* 2018 A serological point-of-care test for the detection of IgG antibodies against Ebola virus in human survivors *ACS Nano.* **12** 63–73

[201] Wu S *et al* 2016 An adenovirus vaccine expressing Ebola virus variant Makona glycoprotein is efficacious in guinea pigs and nonhuman primates *J. Infec. Dis.* **214** S326–32

[202] Ksiazek T G, West C P, Rollin P E, Jahrling P B and Peters C J 1999 ELISA for the detection of antibodies to Ebola viruses *J Infect Dis.* **179** S192–8

[203] Leroy E M *et al* 2000 Diagnosis of Ebola haemorrhagic fever by RT-PCR in an epidemic setting *J Med Virol.* **60** 463–7

[204] Broadhurst M J, Brooks T J G and Pollock N R 2016 Diagnosis of ebola virus disease: past, present and future *Clin. Microb. Rev.* **29** 773–93

[205] Chen Y *et al* 2017 Field-effect transistor biosensor for rapid detection of ebola antigen *Sci. Rep.* **7** 10974

[206] Baca J, Severns V, Lovato D, Branch D and Larson R 2015 Rapid detection of Ebola virus with a reagent-free, point-of-care biosensor *Sensors* **15** 8605–14

[207] Kuno G 2016 Zika virus *Mol. Detect. Hum. Viral Pathog.* **29** 313–20

[208] Petersen L R, Jamieson D J, Powers A M and Honein M A 2016 Zika virus *New Engl. J. Med.* 1552–63

[209] Musso D, Roche C, Nhan T X, Robin E, Teissier A and Cao-Lormeau V M 2015 Detection of Zika virus in saliva *J Clin Virol.* **68** 53–5

[210] Dolai S and Tabib-Azar M 2020 433 MHz Lithium niobate microbalance aptamer-coated whole zika virus sensor with 370 Hz/ng sensitivity *IEEE Sens. J.* **20** 4269–74

[211] Dolai S, Baker B and Tabib-Azar M 2021 Micro fabricated MEMS colorimetric devices to measure Zika-induced residual stress and mass loading *IEEE Sens. J.* **21** 4682–7

[212] Marshall J A and Bruggink L D 2011 The dynamics of norovirus outbreak epidemics: recent insights *Int. J. Environ. Res. Public Health.* **8** 1141–9

[213] Hall A J *et al* 2011 Centers for Disease Control & Prevention (CDC) updated norovirus outbreak management and disease prevention guidelines *Mortal Wkly. Rep. Recomm. Rep.* **60** 1–15

[214] Kapikian A Z, Wyatt R G, Dolin R, Thornhill T S, Kalica A R and Chanock R M 1972 Visualization by immune electron microscopy of a 27-nm particle associated with acute infectious nonbacterial gastroenteritis *J Virol.* **10** 5

[215] Bruggink L D and Marshall J A 2009 Norovirus epidemics are linked to two distinct sets of controlling factors *Int. J. Infect. Dis.* **13** e125–6

[216] Robilotti E, Deresinski S and Pinsky B A 2015 Norovirus *Clin. Microb. Rev.* **28** 134–64

[217] Chiba S *et al* 1979 An outbreak of gastroenteritis associated with calicivirus in an infant home *J. Med. Virol.* **4** 249–54

[218] Rabenau H F, Stürmer M, Buxbaum S, Walczok A, Preiser W and Doerr H W 2003 Laboratory diagnosis of norovirus: which method is the best? *Intervirology.* **46** 232–8

[219] Teunis P F M, Sukhrie F H A, Vennema H, Bogerman J, Beersma M F C and Koopmans M P G 2015 Shedding of norovirus in symptomatic and asymptomatic infections *Epidemiol. Infect.* **143** 1710–7

[220] Chand R and Neethirajan S 2017 Microfluidic platform integrated with graphene-gold nano-composite aptasensor for one-step detection of norovirus *Biosens. Bioelectron.* **98** 47–53

[221] Xiang L, Wang Z, Liu Z, Weigum S E, Yu Q and Chen M Y 2016 Inkjet-printed flexible biosensor based on graphene field effect transistor *IEEE Sens. J.* **16** 8359–64

[222] Poddar S K 2002 Influenza virus types and subtypes detection by single step single tube multiplex reverse transcription-polymerase chain reaction (RT-PCR) and agarose gel electrophoresis *J. Virol. Methods* **99** 63–70

[223] World Health Organization 2011 Manual for the laboratory diagnosis and virological surveillance of influenza *World Heal Organ.* **153**

[224] Metzgar D *et al* 2010 Single assay for simultaneous detection and differential identification of human and avian influenza virus types, subtypes, and emergent variants *PLoS One.* **5** e8995

[225] Peduru Hewa T M, Tannock G A, Mainwaring D E, Harrison S and Fecondo J V 2009 The detection of influenza A and B viruses in clinical specimens using a quartz crystal microbalance *J. Virol. Methods* **162** 14–21

[226] Li D *et al* 2011 A nanobeads amplified QCM immunosensor for the detection of avian influenza virus H5N1 *Biosens. Bioelectron.* **26** 4146–54

[227] Wang R and Li Y 2013 Hydrogel based QCM aptasensor for detection of avian influenza virus *Biosens. Bioelectron.* **42** 148–55

[228] Jiang Y *et al* 2015 SAW sensor for Influenza A virus detection enabled with efficient surface functionalization *Sensors Actuators* B **209** 78–84

[229] Erofeev A S, Gorelkin P V, Kolesov D V, Kiselev G A, Dubrovin E V and Yaminsky I V 2019 Label-free sensitive detection of influenza virus using PZT discs with a synthetic sialylglycopolymer receptor layer *R. Soc. Open Sci.* **6** 1–13

[230] Ilic B, Craighead H G, Krylov S, Senaratne W, Ober C and Neuzil P 2004 Attogram detection using nanoelectromechanical oscillators *J. Appl. Phys.* **95** 3694–703

[231] Migoń D, Wasilewski T and Suchy D 2020 Application of QCM in peptide and protein-based drug product development *Molecules* 25

[232] Zhao X *et al* 2012 Protein functionalized ZnO thin film bulk acoustic resonator as an odorant biosensor *Sensors Actuators* B **163** 242–6

[233] Di Pietrantonio F *et al* 2016 A Shear horizontal surface acoustic wave biosensor for a rapid and specific detection of d-serine *Sensors Actuators* B **226** 1–6

[234] Ji H F, Thundat T, Dabestani R, Brown G M, Britt P F and Bonnesen P V 2001 Ultrasensitive detection of $CrO4^{2-}$ using a microcantilever sensor *Anal. Chem.* **73** 1572–6

www.ingramcontent.com/pod-product-compliance
Ingram Content Group UK Ltd.
Pitfield, Milton Keynes, MK11 3LW, UK
UKHW051852100926
8556IPUK00005B/588

9 780750 338653